NEUROCAPITALISM

Technological Mediation and Vanishing Lines

Giorgio Griziotti

Neurocapitalism. Technological Mediation and Vanishing Lines
Giorgio Griziotti

Foreword by Tiziana Terranova
Translated by Jason Francis McGimsey
English edition edited by Stevphen Shukaitis

ISBN 978-1-57027-342-1

Cover image: Martino Saccani
Cover design by Haduhi Szukis
Interior design by Margaret Killjoy

Released by Minor Compositions 2019
Colchester / New York / Port Watson

Minor Compositions is a series of interventions & provocations drawing from autonomous politics, avant-garde aesthetics, and the revolutions of everyday life.

Minor Compositions is an imprint of Autonomedia
www.minorcompositions.info | minorcompositions@gmail.com

Distributed by Autonomedia
PO Box 568 Williamsburgh Station
Brooklyn, NY 11211

www.autonomedia.org
info@autonomedia.org

CONTENTS

TRANSLATOR'S NOTE

Jason Francis McGimsey

The translation readers will find in *Neurocapitalism* is the fruit of a longstanding collaboration between the author and I. Several of the essays that are now in this book we previous translated for other (mostly digital) publications, either by myself or others. This made the assembly (and updating) of the text somewhat of a challenge, for two main reasons. First, translations age and, over time, once current references quickly become dated and even inaccurate, especially when dealing with digital technologies. Secondly, contemporary political and philosophical terms themselves change quite quickly: while translators race to find the right adaptations in other languages, the concepts themselves change.

Likewise, Europe's social and political context is still very different from that of the Anglophone world and the language used to describe it remains quite "foreign." One very good example is the idea of "precarious workers" (i.e. an employee without a secure contract): on the continent, workers are generally more accustomed to the protections offered by the socialist state, whereas the English-speaking world has had at least thirty years to get used to a more "flexible" labor market. This makes the translation of "precarious labor" or "precariat" sound somewhat mechanical in English, while in Italian or French these terms are a part of daily conversation.

That said, in the end, it is the hopes of any translator that the introduction and use of new terminology, while requiring a small effort on the part of the reader, can in turn enrich our ability to perceive and analyze new nuances in the world around us.

FOREWORD

Tiziana Terranova

In the infinite production of texts, studies and analyses on computer networks and digital media, it is rare to find a book like that of Giorgio Griziotti, which is capable of combining a competent technical viewpoint with a coherent theoretical perspective and an obvious political passion. Griziotti tells us in his preamble how this synthesis was possible, from the moment he chose to implement his subjectivity, following the feminist exhortation that insists and continues to insist (from Donna Haraway, Gayatri Spivak and Sandra Harding to Rosi Braidotti and Karen Barad) on the importance of situated and corporal, partial and partisan knowledge, that unfolds from a specific place and time rather than from a disembodied and impartial perspective. It should be stressed, therefore, that in this text, as Griziotti tells us from the outset and lets us glimpse through all the book, the various existential dimensions in a living search animated by a great political passion nourished by the "common learning" of collective self-training.

Giorgio Griziotti is an IT engineer and programmer: he has therefore intensive hand-to-hand experience with language and codes, an activity which he boldly extends from software programming to test writing. One of the first computer scientists to graduate in the intense cultural, social and political atmosphere of Italy's long and hot 1968 (some say it lasted until 1977), he represents in a sense the Italian variant of the hacker experience told by Steven Levy in his classic history of hackers in the United States. Unlike the latter, its relationship with the computer is not animated by the deeply rooted libertarian spirit of the United States, which expresses itself in the drive to free the computer machine from bureaucratic control. Instead, we must look to the Copernican revolution of Italian Marxist *operaismo*, with its rereading of the "Fragment on the Machines" in Marx's *Grundrisse*, and its discovery of science and technology as the incarnation of the General Intellect: the knowledge embodied in the capitalist productive machine on which the open dance of exploitation, resistance and liberation is played out. Political activism and exile in France, where he worked in a large multinational IT services company for telephony, have led him to personally experiment with both the productive transformations of the General Intellect that mobilize technologies as a means of exploitation, control and extracting added value and the practices of self-training and nomadism to which he entrusts the hopes of new forms of organization. It is in this context that the *operaisti* practice of co-research unfolds in a transformation in which the modes of industrial and post-industrial production are mixed, alternating and overlapping with the corresponding forms of control, resistance and struggle.

Contrary to the classical epistemological politics of the avant-garde aspiring to direct the subordinate classes, Griziotti practiced co-research as a means of producing a transformation of the subjectivity of both the researcher and the places undergone by scientific and technological change. The world of work and production has been reshaped by the omnipresence of new techniques and technologies, such as real time computing and Enterprise Resource Planning, this technique that plans, integrates and subordinates automated manufacturing machinery and cognitive processes for revenue maximization to the enterprise ecosystem. Life itself and consumption have been shown to be biopolitically productive through the integration of free work, emotions, affects and body virtualities into economic valorization. Faced with this, it is a question of valuing the organization animated by an emancipatory impulse that turns technologies against exploitation. Griziotti's writings therefore have a deep affinity with the writings of programmers and hackers, with Richard Stallman's essays on copyleft or with Eric S.'s *The Cathedral and the Bazaar*. Raymond, but also with the Italian texts of Collectif Ippolita or Jaromil, in which the qualified critique of the "state of things" of both computing and computerization always opens up perspectives of liberation and emancipation.

Just as solitary programming activity is transformed by the setting in motion of the common social cooperation of open source and peer-to-peer networks (P2P), the solitary writing of Griziotti's theoretical essay is nourished and transfigured by the participation in self-study modalities activated by eccentric networks and external to the world of official and dominant academic research, with its own independent networks. The learning practice built on the basis of militant cooperation and sharing also differs from the structuring method in terms of both public research modalities (such as that financed by States or transnational bodies such as the European Union on the basis of calls for tender that specify its objectives and methods) and private (with an emphasis on monetizing the results in the form of copyright and patents). The method of co-research underlying this book is in harmony with the hacker spirit, of which Griziotti exposes the limits, but also with the idea of commonfare, or common welfare, on which Andrea Fumagalli and Carlo Vercellone have written. As in *La Cura*, the project of two other magnificent hackers of contemporary knowledge, Salvatore Iaconesi and Oriana Persico, one can fight an existential condition (for Iaconesi to become "cancer-sick" in the dehumanizing machine of hospitals and the national health system; for Griziotti, that of exile and IT engineer in a large company that is de-unionized, precarious and subject to the devastating pressure of internal competition) by sharing and co-producing knowledge that grafts emancipatory dynamics into the division of labour between experts and non-experts, between service providers and simple patients/clients.

Griziotti's participation in the networks of collective political self-education (the seminar "From the public to the common" in Paris and the free universities Uninomade and Effimera) gave him the fundamental thesis that runs through the book: that of the transformation of capitalism, from industrial production to post-industrial and biopolitical production. It is a book full of concepts (biohypermedia, General Intellect, cognitarian aristocracy, biocognitive capitalism, real subsumption), but also of insights into technologies that are as ubiquitous as they are often invisible and opaque (the functioning of computer memory and social media algorithms, customer relationship management, and telephony-computer coupling, ubiquitous applications, video games and so on). We are experiencing a social, economic and technological transition that leads capitalism to move from the industrial stage to the biocognitive stage, to extend production to the heart of life and its forms, in this "biohypermediality" of mobile networks and devices, intimately connected to the body that Griziotti has been documenting for years in essays and articles published on the web, on sites of autonomous university networks or in sociological and cultural journals. This transition is expressed not only in technological transformation and innovation, but also in the way in which technological mediation invests the social body, from allergies and autoimmune diseases to burn-out syndromes caused by hyper-competition between precarity, anxiety and stress caused by the fact that attention becomes the central and integral part of work to the materialization of the mobile phone as a biopolitical device in the mobile phone. Through the three sections (produce, live, organize) and their interludes, Griziotti leads us into the midst of these changes with a vigilant and enlightened eye and an acute perception of the articulation between production, politics and life: the relationship between P2P and client-server architectures; the diffusion of open space and telework; corporate social networking and knowledge management; the relationship between the emergence of robotic intelligence and the political demand for universal income; the environmental challenge posed by the accumulation of electronic waste (e-waste); the cognitarian aristocracy of start-ups; the capitalist integration of crypto-currencies; the new subjectivity of makers and their fablabs. It thus deals with the relationship between communal and financialisation, between the affirmation of the autonomy of the biohypermediated General Intellect and new strategies for capturing value. A complex, articulated and passionate book, full of information, intuitions and concepts, which in itself testifies to the vitality of the General Intellect in the era of contemporary biohypermediatized production.

FOREWORD THE ENGLISH EDITION

Politics, technique and co-research

From a personal point of view, my desire to delve into writing this book comes from the intersection of my two main interests: politics and technology. Throughout my lifetime, these vital passions have been nourished by my practical experience in these fields. My political involvement began with the social movements of the '70s in Italy and, despite a few breaks determined by the somewhat difficult contingencies of political exile, has continued throughout my life while working with different collectives on seminars regarding political analysis and reflection. My passion for technology goes all the way back to my childhood and has manifested itself in an intellectual curiosity and an interest not only for technological innovations but also for the sociopolitical implications they bring. My studies therefore naturally oriented themselves towards the technical domain, up to obtaining my degree in engineering with a specialization in the field that would later be called Information and Communication Technologies (ICT*[1]) whose development has never slowed over the last few decades.

1 From here on, an asterisk* indicates a reference to the Glossary at the end of the book for more details. The acronym ICT will be used throughout the book.

The ICT of the '70s didn't benefit from the dominant position it has today and, at the time, it didn't seem very important to try to understand how the technological innovations of capitalism could turn into an antagonism that radically transforms the present. However, having posed this question relatively early, I found myself in the position of a technician who walks, a bit unconsciously, along the path of co-research.* At its roots, co-research is a practice born from activist field research with the workers of the Fiat Mirafiori and other Piedmont factories, including Engineer Adriano's mythical tech company, Olivetti. There, the work of Romano Alquati, one of the founding fathers of this practice, is well noted. In his own words:

> an activity that is both research and process of consciousness as well as a reciprocal transformation of the identity of the researcher and that which was in those years being called worker subjectivity. Co-research, for its egalitarian character, is counterpoised and substitutes the old avant-garde communist practice of guiding the masses in the struggle.[2]

It seems useful to detail – in chronological order – the three types of experience that have characterized my personal relationship with technological innovation in order to situate the ideas and the hypotheses developed in this book in a social and political context.

The first experience was in the context of my years as a consultant for large software applications in an IT consulting multinational corporation. At the moment when capitalist innovation began extracting knowledge from third-sector workers in order to codify it into software packages that would then guarantee company control and the optimization of rent, the process of co-research takes on a new dimension. With IT consultants on one side and the "investigated" cognitive workers on the other, these two groups are usually set against one another by management. The whole challenge in practicing co-research lies in creating a space where, instead of competing, the cognitive worker can gain consciousness of the company's financial and control objectives thanks to the automatization of internal procedures and, later, acquire the capacity to redirect, modify or even

2 Armano & Sacchetto, 2012 [our translation].

block the processes of digitalized exploitation. An arduous mission in an era when union influence was being visibly reduced and when labor organizations were entering the phase of destabilization and decline that is now quite evident. An early signal of the hostilities that were manifested in companies were the articles published in specialized technical journals warning managers of the risk of the failure of software projects to control workers through automation: it appears, in fact, that only a third of them were successful. Even though only technical explanations were given, it was pretty clear that the forms of cognitive worker resistance made the road to company normalization through dangerous technology.

The second type of experience is more fragmented, precisely due to the political activity that provoked ruptures and forced me to start all over again – multiple times and in geographically diverse places – with my attempt to spread knowledge and alternative/antagonistic uses of technological innovation. Whether it be creating a cooperative start-up in Milan in the early '70s or, in the early '80s, an association working to promote the first PCs education in non-profit social economies in France, or even in conceiving the first network application of labor rights to support union activities and the use of free software, there has always been a thread tying these experiences together; experiences that, in a certain sense, come close to that of the modern world of hackers.

The third and last type of activity is tied to the intuition of the impact that the spread of networks and mobile technologies has on society. This intuition gave me the opportunity to set up an experimental laboratory to implement the very first mobile applications, the famous "apps" that will later be discussed at length.

The Cathedral and the Bazaar

At the beginning of the digital era, a certain similarity still existed between writing an article or a book and writing a program, despite the differences between natural languages and software code. In both cases, however, it was a predominantly solitary activity for which a long and tiresome testing phase was needed before the program "ran" or before the writing was fluid. In the first case, objectivity prevailed over the author's subjectivity. With the spread of networks and the birth of new movements, especially the hacker movement, modalities of cooperation changed and, consequently, the method for writing software was profoundly modified. This was made possible by a timely integration of a global community. The programmer's original solitude is tempered by this potential. Making contributions available and being able to cooperate in an open context, participating in various projects, introducing evolutions or simply correcting errors become the commons* of free software.* In a noted article, E. Raymond[3] uses the metaphor of the "bazaar"[4] to evoke a methodology that lacks hierarchy in the common work of hackers which enabled the success of Linux,[5] an extremely complex artifact. He counterpoises this way of cooperating to the "cathedral," i.e. a restricted nucleus of experts and specialists who construct a masterpiece; a technique that, according to the author, doesn't lend itself to creating great works of software. The "bazaar" method has had an influence on writing this book, both for the intensity and the frequency of my exchanges on the various topics addressed, as well as for the support that many people have given me to compensate for my numerous weaknesses, including the literary and linguistic shortcomings of an engineer who emigrated decades ago.

Another part of this "bazaar" attitude can be seen in my potentially ingenuous attempt to make the materials accumulated from numerous articles, texts and books (in various languages) available in the form of a digital library. Unfortunately, this initiative was blocked due to copyrights infringements using the Digital Millennium Copyright Act (DMCA), surprisingly coming also from authors and publishers known for their sympathy for copyleft principles.

While it is difficult that the diffusion of knowledge is stigmatized or prosecuted in the hacker movement, the same cannot be said for the academic and publishing world where this principle is less valid.

---

3 (Raymond, 2001)

4 E Raymond is also a fierce opponent of Copyleft and of free software, against which he proposes open source, a term of his invention. In this regard, see (Ippolita, *Open non è Free. Comunità digitali tra etica hacker e mercato globale*, 2005).

5 Linux is the family of Unix-like operating systems, released with the GNU GPL licenses that characterize free software (see the Glossary), under various possible distributions, with the common characteristic of using Linux kernel as the core. For more details, see Linux in the Glossary and the later paragraphs dedicated to Unix and Linux.

Even though the great majority of what is published in my techno-socio-political digital library is in relation to the ethics and the practice of free software, the copyright machinery has not stopped and continues to impede the free circulation of knowledge.

INTRODUCTION

Nobody liberates anybody, nobody liberates themselves, people are liberated through the mediation with the world.
(Freire, 2002)

Marxian Cyborgs

Technological mediation is inseparable from the origins and the history of humanity in that it is an integral part of human mediation with the world and therefore society. This book is dedicated to the invasive role of such mediation in contemporary cultural, anthropological, social, political and economic dynamics that teeters between servitude and autonomy. There has been no time when technique was not an essential part of human activities and did not condition human life. The extreme hypotheses of this affirmation can be summarized in two scenes by cult filmmakers and writers like Stanley Kubrick, Ridley Scott, Arthur C. Clarke and Philip K. Dick. First, the famous sequence in *2001: A Space Odyssey*, when an anthropomorphized monkey, inspired by the discovery of a 4000-year-old monolith, picks up a bone to use as a weapon for the first time. The second is the ending of the famous monologue that the android Roy Batty recites dying under the torrential rain in the final scene of *Blade Runner*: "I have seen things that you humans cannot even imagine…"

In a relationship of reciprocal exchange, machines have always formed subjectivities and vice versa. In the still recent industrial era, the automobile and its variants assumed the essential function of mediation with space-time. In the collective imagination of that era, the conjunction of man-machine inspired hybrid entities, correlated with mythological or astronomical figures like the centaur and the racing car. This went so far as to become the central subject of one of the cult novels of that period, *On the Road*, in which the famous scroll of teletype paper on which Jack Kerouac[1] wrote in one breath in 1951 is transformed into a symbolic prolongation of the strip of asphalt travelled at the wheel.

From a political point of view, the way in which the subjectivity of the mass worker was forged, notably in response to the forms of capitalist organization of Fordist assembly line production, is even more incisive. We could continue on with the telephone, the radio or the television and the infinite examples from each era, but our central hypothesis is that contemporary technological mediation is of another order, because it has become the hinge to a veritable anthropocenic passage, i.e. a geological era in which human activity has a global influence on the terrestrial ecosystem. Some, quite rightly, prefer the concept of capitalocene based on the fact that capitalism is the main culprit of current environmental imbalances. We might even have to speak of postanthropocenism, where the epicenter is occupied by posthuman figures that constitute the fusion of technosciences and matter, including living matter. Figures that, from Donna Haraway's cyborg[2] onwards, emerge more and more often in the sociological, philosophical and cultural debate, but with biopower's constant instrumental and political use of technology.

Maybe we have reached a point of inflection where multiple factors contribute to the unstoppable decline of Humanism – in its predominantly Eurocentric, white and male nature, inherited from the Enlightenment – even in its socialist forms. This decline started back in the '60s and '70s, when the extended social movements of the time, although failing in their central intent to politically overthrow a capitalism that instead was reinforced, created irreversible normative fractures from which diversified and hybrid singularities emerged. The cycle of feminist, ecological, antiracists, anti-imperialist and anticolonial struggles, as well as those for sexual liberation, broke with classic humanist dogma and which the opposition defined as anti-humanist.

1 Kerouac, 2010.

2 Haraway, 1995.

Today's technological mediation is no longer limited to exercising growing influence, but rather integrates matter in the whole of vital processes that go from genetic hardware to affective software. It is this tendency at the base of posthuman and transhuman theories to treat it as a change in the paradigm of great contemporary importance. While Transhumanism has embarked upon a perverse and pathological path towards technology that would make us omnipotent and immortal, the hypothesis of posthumanism as the definitive surpassing of Humanism, as an articulation and prolonging of anti-humanism, is doubtless fascinating. However, we cannot forget the domination that neoliberal regulation has exercised for nearly forty years now, and that has generated, among others, contemporary forms of submission to the power of death, according to the "necropolitical" definition provided by Achille Mbembe.[3] A control that, having engulfed reformist social democracies, appears more threatened by the consequences of its own politics (like the impoverishment of many populations, wars and, above all, widespread ecological degradation) than by the cyclical surges of social movements.

In this context, there is no guarantee and even fewer truly tangible indications that "posthuman postanthropocentrism"[4] can alone assure the emergence of an ethics that avoids the difficult future we are destined for if we continue in a direction dominated by neoliberal economic rationality. To avoid losing ourselves to excessive discouragement or by an equally unjustified optimism, we suggest examining the process of the formation of new subjectivities, not only in the context of technological acceleration, but also and *in primis* in the context of contemporary capitalism: biocognitive capitalism.[*] Other questions emerge for which we will try to provide some answers: in what way can mediations operated through biohypermediactic and biogenetic technologies be used to harness or liberate the energies of the social body, and by whom? What kind of subjectivities will they generate? We are ever further from the dichotomies of the industrial era when it was easy to put the means of production of fixed capital on one side and, on the other, variable capital and living labor. Even if we can agree that a part of fixed capital is now constituted by generalized technical know-how, it remains to be seen how technological

3 Mbembe, 2014.

4 Braidotti R., 2014, 63.

mediation generates hybrid subjectivities when old Marxist concepts are difficult to apply.

Here another question arises: do we have the adequate conceptual tools today to analyze and politically act in this context? For example: can the Marxist categories of formal and real subsumption function in the same way in a world where subjects, procedures, relations and contents change nature?

The Three Parts: Producing, Living and Organizing

The choice to structure this book according to three main aspects is functional to an analysis of how technosciences operate within each of them, producing different effects and interactions. In examining the variegated spectrum of technological mediations, diverse registers in which they are inscribed appear: biopolitical, chronological and cartographic. Registers that are integrated and entwined with existential aspects of production, life and organization. We have furthermore tried to consider the shift in the perception of time and space in following the transformations that are underway. How can we not think about the fact that every entity that is connected to a network is geolocalized and transmits a pronounced time from its internal biological or artificial clock? This subdivision also poses the problem of how to delimit the sections when instead the separation between labor and life is increasingly unstable and confused, resulting in the passage from an industrial worker's "producing to live" to today's precarious cognitive worker "living to produce." With the intention of allowing alternatives to reading this book sequentially, we have also tried to give each section its own autonomy in structure and comprehension.

The first section is dedicated to "Producing," where the most consistent ties remain related to the recent past of industrial production that, although it has lost its central role, still endures today. We are still very much influenced by this period and by the notable presence of the tutelary deity of that era, Karl Marx, who allows us to begin asking the questions necessary for building a framework for biocognitive capitalism: how should we treat the weight of technology in cognitive production without putting the machines from the era of carbon and steel on the scales? How can we understand contemporary accelerationism without evoking the birth and implementation of the concept of real time computing born more than sixty years ago?

How can we comprehend the digital native's* desire to escape without having lived through the stress of corrupted cooperation of capitalism where competition and fear are omnipresent?

We have dedicated the second section to "Living." Contrary to the modes of production that haven't broken all ties to the past, bios and zoë* – i.e. conscious life and undifferentiated living matter – move and are directed towards new dimensions through contradictory and divergent uses of technology. Among other things, the sphere where the process of blurring between living and working brings with it the seeds of the separation between income and labor. Interrogating "living" comes from the search for points of reference within this universe: how can we explore the metamorphoses of subjectivities put through multiple procedures of normalization without observing the signs inflicted or revealed by our bodies? Which perceptive entities are affected by diffused technologies thus shifting the framework of life? What are the consequences on subjectivity?

The term "Organizing," the title of the third section, is possibly even more generic than the two preceding: changes in social behaviors is evident in this field. One could ask if this appearance is not accentuated by the deliquescence of the historical forms of organization of the working class that, in Europe and elsewhere, has often passed from being an antagonist to being a political liaison with financial governance. Where can we begin in this situation? We have chosen as a starting point the comparison between the lifestyles of these two eras: on one side, that of industrial capitalism, characterized by belonging and, on the other, contemporary capitalism, characterized by migration and nomadism and where ICT* have brought the speed of hypermediatic ties to the heart of the matter. From here, we face the questions regarding the impulse to migrate toward spaces that the system tries to fragment, delimit or simply render inaccessible (like the dramatic circumstances of refugees and migrants who attempt to reach Europe by any means and risking their very lives). A separate reflection is dedicated to the antagonistic relation between technical and religious thought. Developing the writings of Simondon on this theme, a surprising relationship between the new paradigm of technological mediation and the expansion of archaic and dogmatic religious forms emerges.

Two "evasions" function as intermissions between the three sections and a third at the end. The first, between "Producing" and

"Living," evokes the hypothesis of a *space devouring time.* Once the machine is able to interact in a "living" way with "real" time, it becomes a part of the territory and devours it. The second evasion closes the not-so-virtuous cycle of new postindustrial production. "E-Waste" is the term used to designate the technological dumps concentrated in "peripheral" countries that, in the capitalist subdivision, are garbage sites for the planned obsolescence of technological objects. The final evasion is dedicated to Castel del Monte, a site that shares the foundations of digital technologies in that its architecture is entirely based on octaves, recalling the byte, the unit of eight bits used to encode a single character of text in a computer. This monument has expressed for nearly eight centuries a conception of science and technology that opposes today's dominant vision because it articulates the beauty of the notions that understand it as a work that seemingly avoids any finalism. Produced by a collective, the Court of Federico II, where heterogeneous cultures were represented, it has become an essential reference that allows us to understand how the power and the spread of contemporary technologies can become a key tool in the construction of the Common.*

Part I:
PRODUCING

In our collective imagination, the verb "to produce" is tied to industrial labor and evokes terms like "production," "productivity" and "GDP." Here we extend the meaning beyond material production to include cognitive production and "man-to-man" production. In this context, affects become an essential component of knowledge and skills, thus we have privileged "to produce" over "to work" or "to create." However, "to work" harks back to the dualism that opposes life to labor, one that is less and less real because both are being diluted into a continuum. Meanwhile, "to create" is too selective because it only appeals to creative labor. In examining and analyzing "production" from the angle of technological mediation, we have tried to avoid any deterministic approach despite the temptations generated by the current accelerations.

Deconstructing industrial production towards the digital production of General Intellect

Notwithstanding the metaphors that refer to the industrial lexicon used in large digital projects,[1] we still are not used to thinking about

1 In digital projects conducted by IT service companies, the activation of a new software application is often referred to as being "put into production." Moreover, industrial terms are often used in project management. Monthly production is,

cognitive services as "production," a term that remains heavily associated with the previous industrial era. Along the same lines, we are inclined to relate today's tools more with the knowledge economy and with cognitive labor than with materiality. The adoption of and the ability to use mass technologies, like the internet and mobile communications, are born above all in personal and private uses and are only later, with a notable delay, transferred to the professional sphere. We therefore tend to underestimate the impact of ICT on all production and especially on material and industrial production. This is why we are proposing a pathway that occasionally touches on concepts and innovations that have also transformed material activities. We have highlighted five of them that preceded and contributed to creating today's context, purposefully excluding from this list the internet, which is the unifying matrix of all ICT and the core of the knowledge economy, precisely because it constitutes the very theme of our analysis. The roles of a handful of these innovations are well known and have been amply analyzed, while in other cases they have remained in the shadows. They are:

- Unix and TCP-IP
- Interrupt and real time computing (RTC)
- Graphical User Interface (GUI)
- Personal Computer (PC)
- Applications of Enterprise Resource Planning (ERP)

We intend to stress how the genesis of these innovations was profoundly different from those of industrial technologies and how this has consequently provoked an acceleration in the mutations of technological mediation.

Another observation along our "production" pathway elevates collective intelligence to the role of protagonist in this revolution. Known in Marxist terms as "General Intellect," this concept becomes ever more important due to the geometric acceleration of new technologies. Beyond the profound changes to subjectivity induced by the use of mass technologies, here we are interested in evoking the dynamics of interexchange with collective intelligence. Dynamics that are often contradictory because these innovations also lend themselves as tools

for example, the amount of days spent by a team to develop software, etc.

to extract value and expand control. Quite often, these two "opposed" uses are integrated and confusing, making any analysis much more complex. The more "progressive" wing of the system has the intuition that, in order to increase levels of rent, they must leverage a type of cooperation that is very different from the kind found in a Fordist factory. The cognitive factory is the metropolis and, in the space-time of the metropolis, both the machines and the workers are different.

In order to analyze the deviational risks inherent in cognitive production today, this section ends with an investigation into the labor conditions in IT and software consulting services.

INNOVATION AND GENERAL INTELLECT

The term "innovation" is an ancient one that refers back to the Latin *novus*, or the Greek *neo*. The fact that it is now often accompanied by the adjective "technological" only confirms today's presumed hegemony of technosciences. On Wikipedia, "innovation" is the application of an invention or discovery. The third edition of the Oslo Manual of the Organization for Economic Co-operation and Development (OECD) defines it as:

> Technological product and process (TPP) innovations implemented technologically new products and processes and significant technological improvements in products and processes. A TPP innovation has been implemented if it has been introduced on the market (product innovation) or used within a production process (process innovation). TPP innovations involve a series of scientific, technological, organisational, financial and commercial activities.[1]

1 See: OECD, *Oslo Manual Guidelines for Collecting and Interpreting Innovation Data Third Edition*, 2005, 25.

This definition reveals that today innovation no longer comes only from Research and Development laboratories (R&D) as it once did in the industrial era, but that it is rather a product or a process where technological invention is integrated into a world where the multitudes are limited to the categories of consumers and human resources; categories that tend to blur.

From the Ford Model T to...

In a more recent edition of the same manual, the OECD was forced to recognize that, today, a large part of innovation no longer relies on R&D laboratories but instead on various factors among which worker know-how is primary:

> Knowledge and technology have become increasingly complex, raising the importance of links between firms and other organisations as a way to acquire specialised knowledge. [...] While R&D plays a vital role in the innovation process, much innovation activity is not R&D-based, yet relies on highly skilled workers, on interactions with other firms and public research institutions, and on an organisational structure that is conducive to learning and exploiting knowledge.[2]

Even more recently, the same organization has even began to support crowdsourcing initiatives.[3]

Historically it hasn't always been this way. If we go back to the last decades of the 19th century, i.e. at the beginning of the second industrial revolution – characterized by the diffusion of electricity – we find that it is precisely at this time that the first R&D labs in the German chemical industry appear, to spread later to other industrial branches. The innovations of this phase – Fordist production based on the invention of the assembly line, the spread of the first communication devices such as the telegraph and the telephone, or certain extraordinary discoveries in chemistry – all had a common denominator in

2 OECD, 2005, 29.

3 See, for example, *Trusting in crowds*, http://www.oecd.org/about/publishing/oecdobserver/trusting-in-crowds.htm.

their genesis. Beginning with broader scientific knowledge that was, however, still limited to an elite, the phases of research and the detailed description that made invention reproducible were often confined to circumscribed and specialized environments like, for example, dedicated R&D laboratories.

At the same time, the transformation of the prototype into the process of mass production that took advantage of large economies could not take place without a move to subject laborers to the assembly line. Marx defines this as "real subsumption" because workers are expropriated of a large part of their know-how, then concentrated in machines and, in the specific case of the automobile, in the assembly line. Workers could not therefore do anything more than submit to fragmented labor processes. Among the factors that were a part of this phase, we find the emergence of a middle class that becomes the "potential market" where a massive quantity of products are unloaded. The most symbolic of examples is the Ford T, "the car that put America on wheels." Between 1908 and 1927, more than fifteen million Ford Ts were produced and, over this long period, very few major changes were available, explaining Henry Ford's famous joke about his clients' freedom: "Any customer can have a car painted any color that he wants so long as it is black."

This reduced ability to choose corresponds to the limited contact between the producer and the consumer in the long lifecycle of a product. Robustness and durability were the paradigms of the industrial period. The manufacturer saw client or user assistance as a source of cost rather than profit. Innovation is introduced into the market after a long series of passages that stray from the conception of industrialization. The latter implies the creation of complex machineries for processes of commercialization and distribution networks. It follows that the interval between the idea or the initial discovery and its widespread use was quite long: sometimes years, or even decades, passed. At this time, the initial invention and the prototypes of products, and later services, were often protected and kept secret. The same can be said for the life of the product, as in the case of the Model T. Even if "technological innovation is not a phenomenon of isolation[4] "and reflects a state of consciousness and scientific knowledge, the process is limited to an institutional and particularly industrial environment.

4 Castells M., 2008, 37.

Potential users of the Ford T were not aware of the existence of the product neither during its conception nor in the testing or integration phase, and this barrier of separation between designers and users persisted throughout its lifecycle: the first were "learning by doing," while everyone else was "learning by using." In the '70s, the mass worker became the mass consumer of the goods he produced, like the case of FIAT cars in Italy – the Seicento and the Cinquecento – that granted a higher grade of mobility. However, by no means does this imply that the industrial worker contributed in any way to designing, conceiving or test driving it; workers were rather a new segment of the market for this new product, and this is precisely the barrier that was destined to collapse in the IT revolution.

This situation corresponds to the capital/labor relation at the time, characterized by the separation between intellectual and manual labor and the concentration of knowledge. Evolution was slow: one must first produce and sell great quantities of identical products, consequentially the products had to be relatively accessible to new social classes. The creation and development of mass consumer credit thus played a complementary and fundamental role. On one side lie great investments – implying the strategic role of fixed capital and a high-volume, rigid production process in order to maximize profit. On the other, we find the repetitive labor of the assembly line and, in general, the fragmentation of mass worker tasks. Moreover, growth in productivity was obtained through the exploitation of economies, static in scale or dimension. The stability of the capital/labor relation was assured through a dynamic that tied workers' salaries to earnings by maintaining the availability of a disciplined workforce in the context of social order.

At a certain point, this dynamic hit its limit. Struggles in a phase of expansion would, a few decades later, change the world. Workers' salaries increased and in the thirty years following WWII, essential social achievements – mediated in Europe by social democracy – would be made, such as generalized access to education and healthcare. The creative capacity of a society with a high cultural potential sprang forth from the rise of exponential development of ICT tools that would then breathe life into new and diffused technical knowledge. It is in such a context that the symbolic product of this emerging epoch no longer emanated from the assembly line but rather from an artifact that was infinitely reproducible at no cost. An immensely complex, immaterial good that is called Unix.

...LINUX: a giant leap forward

The dispositive artifact that, more than any other, represents and symbolizes the transition towards the digital cognitive age is probably Unix. Unix is an Operating System (OS), the system software that manages computer hardware (e.g. the processor and all the input/output peripheral devices) and software resources and provides common services for computer programs. Usually an OS is one of the most complex and delicate software applications to perfect. Unix was conceived in 1969 in one of the largest technological laboratories of the industrial era, AT&T's Bell Labs, the giant telephone operator which, at the time, held a monopoly in the United States.

When Ken Thompson and Dennis Ritchie, researchers at Bell Labs, started working on and building Unix and the programming language C, they didn't have any commercial objectives. Furthermore, AT&T would be forced by the United States government to release the results of Bell Lab's research in 1974. Among other things, this entailed the obligation to hand over the Unix license and its source code to universities for a nominal fee. In the middle of the Cold War ARPANET, the network protocol which was to be the predecessor of the internet, was financed by the military with the aim of establishing a network that would remain active even after a nuclear attack. However, the military does not have adequate research structures and, consequently, this task is delegated to universities that, beginning in 1980, can utilize an operative network with the famous TCP-IP protocol still in function today.

The University of Berkeley, cradle of the social movements of the '60s, participated in this project. There, a technical team, attracted by the access to and availability of the Unix source code, created a version that integrated the TCP-IP communication protocol into the OS: the famous Unix BSD, the same that still exists today, was born. The team even created a new type of free software license for it. The connection between the claims of the opposition movement that culminated in the '70s and the search for free spaces and innovative modes of communication appears significant and anything but random. Protests against the Vietnam War explode across American universities and Berkeley is at the forefront of the movement. It comes as no surprise that a free code for developing software is created, free from copyright control, and that new technologies and communication protocols become tools for free thought and peer to peer cooperation. A virtual community of postmodern artisans who construct a new mode of production is created: the first community of hackers.

Observed from another angle, this is also a meeting of the libertarian desires of the movements symbolically represented by May '68 against war, but also against sexism, for a counter-culture, where the "conception of knowledge [was that] theorized at the time by the founders of the economic theory of knowledge and the sociology of science, Kenneth J. Arrow and Robert K. Merton respectively." For Arrow, "unlike material goods, knowledge is not destroyed by consumption. On the contrary, it is enriched when it flows freely between individuals. Each new knowledge generates another knowledge according to a virtuous circle that allows each creator, as Newton reminded us, to be like 'dwarfs on the shoulders of giants'." Merton "shares and complements this representation by defining the ethos of science and the rules for regulating the public research activities of scientists according to the principles of open science," where collaboration and openness and the rejection of commodification are at the heart of scientific practice, while interdisciplinarity and open exchange are considered as engines of science and innovation.

This process reveals and confirms the transition to a phase in which the role that capitalism had played in the industrial era became antiquated: the conception of even the most complex products, as is the case of Unix, not only no longer requires investment in expensive machinery but also, and above all, can be accomplished without recourse to the modes of organization and retribution typical to corporations. In this ebullient context, Richard Stallman,[5] a brilliant artificial intelligence programmer at MIT, founded the Free Software Foundation (FSF) in 1984, an association that still exists today and that, among other things, established new legal rules and the GNU GPL or GPL license, currently applied to about two-thirds of free software.

The FSF was possibly one of the first forms of organization for this community and came just after the tensions that brought on the first court battles – like the one between the AT&T Corporation and the Berkeley group – to make the Unix source code open and free. Stallman needed a modern and functional operating system like Unix in order to develop free software, but the latter couldn't be "proprietary software".[*] He thus launched the GNU project, the famous recursive

5 Stallman, 2003-2004.

acronym for "Gnu's Not Unix," underlining both the affinity to and the difference from Unix. GNU's objective is to furnish alternative free software in response to the growing risk of the multiplication of proprietary versions of Unix[6], developed by large computer manufacturers like IBM (AIX), Hewlett Packard (HP-UX), Sun Microsystems (Solaris) or niche software actors like Santa Cruz Operation that, with Xenix, launched one of the first versions of Unix for PCs[7].

Curiously, at first Stallman wasn't immediately able to realize the fact that, beyond the validity of the copyleft[8] principles that he invented through his GPL license, the OS' real political strength lies in the productive power generated in common cooperation. The first draft of GNU is an ambitious project to construct a "cathedral"[9] through the participation of a handful of experts. Stallman didn't have the instinct to solicit, involve and take advantage of the contributions from the hacker community that had formed around the operating system and to which he himself had made a fundamental contribution with the creation of the FSF. We would have to wait another

---

6 Proprietary versions are, in a large majority, developed either starting from Unix BSD, the previously available source code, or from the "System V" version, property of AT&T and from which large companies like IBM and H&P acquired the license. System V also served as the inspiration for creating and normalizing the interface denominated POSIX (by Richard Stallman) that aimed to make the various versions of Unix compatible and which would later be adopted by Linux.

7 Despite the fact that these versions of Unix were proprietary, the advantage for users or designers of application software lied in not being held prisoner by a system or a particular hardware manufacturer as before. Within the Unix family, packages could be easily moved from one system to the next. Finally, it is import ant to note that the second largest manufacturer in the world after IBM was the Digital Equipment Corporation (DEC) and they obstinately refused to develop an official version of Unix for its minicomputers (even if an unofficial one, Ultrix, existed) that ran with proprietary operating systems in order to maintain profit margins. The latter would then decline rapidly and DEC was absorbed by the now-defunct Texas PC company Compaq.

8 "Copyleft" designates a copyright management model based on a system of licenses through which the author (in as much as the original holder of rights over the work) indicates to users that the product can be used, diffused and often freely modified, as long as a few essential conditions are respected.

9 See "The Cathedral and the Bazar" in the Foreword.

generation for someone to make the great GNU project enter into the era of the "bazaar." In fact, a few years later, a twenty-two-year-old Finnish student, Linus Torvalds, unsatisfied with the didactic system Minix, would try. He rewrote a Unix-like kernel[10] that he put online in 1991 so that others could contribute to its development. It would be named Linux (GNU-Linux) in honor of its creator and licensed with Stallman's GPL.

More than twenty years later, thanks to this principle of cooperation in hacker communities, Linux, along with other versions like Unix BSD, under non-restrictive licenses, are still the key elements of our technological-cognitive era. Linux is omnipresent in our lives, directly or through derivative forms, with versions embedded in all kinds of different tools and devices. There are systems that are expressly adapted to various devices like smartphones, tablets, readers, residential gateways, video game consoles, automobiles, network routers, smart televisions and a myriad of other devices that will only multiply in the near future. There are versions or distributions of Linux that run 90% of the supercomputers in the world, including the ten fastest, but it is also the system found in the Raspberry Pi, a truly educational device available for $30 US dollars. Linux and Unix run over 70% of the servers in the world.

Furthermore, incompatible derivatives from free versions of Unix continue to exist; yesterday it was IBM and H&P, today the champion of closed and proprietary systems is called Apple, with its iPhone/iPad Operating System (iOS) and Mac OSX, both derived from Unix BSD. These are examples of how common production becomes, in economic terms, a positive externality that aliments biocognitive capitalism. For Eric S. Raymond, open source theorist and enemy of free software, open source is first and foremost a software development model that assumes that with a large number of developers capable of looking at source code, errors will become obvious. He sums it up in the expression "given enough eyeballs, all bugs are shallow." Free software is above all a philosophy of freedom: the right to modify and use code produced by others creates a virtuous spiral and guarantees that data stored in software will remain available to society. Among

10 In digital technology, the kernel constitutes the central nucleus of an operating system and, in this case, Linux's kernel was created by Torvalds based on the kernel found in Unix.

the supporters of free software, some also believe that it is necessary to adopt copyleft logic, i.e. that it is obligatory to give back to the community, and that what is built on free software cannot be privatized. Others, on the contrary, are not concerned with the second life of their software, following the logic of many universities, such as BSD (Berkeley) or MIT licenses. Raymond's position is also criticized by heterodox economists:

According to Raymond, having taken into account the fact that the development of free software is more efficient than owned software, the market economy would already carry out all the work of copyleft without discouraging new entrants to the market. In fact, it is a position that distorts the spirit of the commons of free software and proposes to back their absorption inside a new business model of the large enterprises in the IT sector.[11]

The Ippolita collective has also contributed in legally and politically distinguishing between free software and open source.[12]

Unix / LINUX: engines of the knowledge economy

Today, when the term Unix or Linux is used, most of the time it is done in a generic fashion or to designate a Unix-like operating system that in form, execution and style belongs to a Unix or Linux genealogical family tree. In order to appreciate Unix's contribution better, we have to take a step back and return to the base idea of the first operating systems for computers, from the tube-based Eniac all the way to IBM's widely distributed mainframe.[13] From the very beginning of the marketing of the first electronic computers, all the way up to the '70s, such a conception only followed industrial-era thought where new concepts, knowledge and inventions applied to production and distribution and were incorporated into a monolithic product. But the conception of the computer, based on the separation of hardware and software, made it so that innovation was no longer exclusively cabled into materials like mechanical machines developed for determined, limited and unchangeable functions. The computer

11 Vercellone et al, *Managing the commons in the knowledge economy*, 2015, 83.

12 Ippolita, 2005.

13 Mainframe or mainframe computer is a type of centralized computer characterized by high processing speeds.

is, by definition, a modifiable and programmable machine, even if in the initial phase computer manufacturers more or less artificially held a monopoly over software in general and operating systems in particular. For this reason, systems were defined as proprietary.

First-generation computers allowed programmers to code specific programs, but the entire development environment – from the operating system to programming languages, i.e. all the tools necessary for developing and running new programs – was only supplied through licenses or expensive monthly leasing fees. In addition, companies exclusively supplied binary executable versions of tools so they could not be modified by external programmers. In this case, the software has the same statute and the same function as hardware: it is developed and sold in binary code. If it does not work, only the maker can be asked to correct the anomaly. Anecdotally, it seems that it was initially a bug[14] in a proprietary and unmodifiable binary program that impeded the correct functioning of one of Xerox's first laser printers that gave Richard Stallman the idea that would lead to the creation of the Free Software Foundation.

First Unix and later Linux characterize the innovation of the cognitive era precisely because they break the chain that prevents the free use of computers. The creators of Unix, in fact, didn't try to build a stronger or faster machine but, by separating the conception of the operating system and program development environments from the hardware it was destined for, they created an alternative that put an end to the control exercised by large computer manufacturers over writing software and over the modes of valorization of labor in this field. Up until this time, it was common for company clients who wanted to change supplier and pass from IBM to their competitor Honeywell to be forced to change not only the materials and software packages they had been using, but also face a heavy series of expensive inconveniences like having to reinsert earlier data or adapt themselves to new application functions[15]. With the arrival of Unix, this form of application monopoly broke down and large computer manufacturers

---

14 "Bug" is a term used to designate an anomaly in a computer or program. Although this originates in 1947 when an insect found in the circuit board of a computer had caused a malfunction, the term had already been used by Edison in 1878 to describe small technical issues in his inventions.

15 For example: managing orders, accountability, pay, etc.

thus lost the exclusivity, price control and evolution of final applications. New prospects opened for developing innovative software, freed from the yoke of proprietary systems, now independent from the type of computer or any specific manufacturer and consequentially becoming ever more accessible.

Unix, therefore, has the characteristic of being the first evolved multi-user, multi-task[16] operating system that, with the C language, constitutes a software environment ideated for use with all types of computers. This is why it is called "portable," breaking through the barriers created by so-called proprietary operating systems that oblige programmers to use specific languages for a specific computer or family of computers.

Instead, Unix concretely puts into practice the separation of code and data from the machine that runs them, and for this it is quite possibly the best symbol of the transition into a new era. With the separation introduced by Unix, machines can be used for infinite functions and the application software that makes this possible can be reproduced in unlimited quantities at minimal costs. Unix and Linux are artifacts of living labor, a far-reaching collective endeavor in which the foundations of the digital age are incarnated: every digital or technological device that contains intelligence uses ideas, concepts and often the code from one of the many branches of this genealogical tree. Unix, the "rebel child of '69,"[17] is the technological product, but also political project, that best represents the generation that started the largest movement in the second half of the "short century," as defined by the historian Eric Hobsbawn as extending from 1914 to 1991. This movement successfully opposed the imperial wars of the time, managed to change cultural and political paradigms and, later inspired by Linux and free software, would characterize the aspirations of the next generations. Both systems acted as catalyzers for a movement of researchers, students, programmers and engineers – not to mention expert users – who broke from the dogma of the privatization of technological knowledge through the establishment of intellectual property rights. The objective of corporations, in the era preceding the personal computer, was to make the highest possible profit

16 An operating system with support for multitasking allows computers to run more than one program at a time.

17 Gubitosa, 2007, 183.

margin. This meant making expensive and inaccessible new cognitive tools such as computers, software and networks, when instead the cost of realizing the latter was spiraling. This generation refused and fought against the very principles of this new subjugation. Free cooperation between peers is at the base of the birth of the hacker movement, a movement that today is still expanding and evolving.

Unix/Linux, like all free and open production that followed, are adapted, improved and made to evolve by a collective that constitutes one of the most significant expressions of the common. Inside this movement, new prospects are constantly renewed for users and the frequency of new versions and new applications increases, whether we are talking about open source versions for free software distributions or programs under commercial licenses. Unix, Linux and all the other versions of this family continue to be in full expansion more than forty years after their original conception and, even if it is difficult to calculate accurately, power billions of devices. Today, proprietary systems and software continue to exist and are widely diffused: the best-known examples are those of Microsoft and Apple, which we will discuss in the next section.

Apple: innovation according to biocognitive capitalism

Steve Jobs praised Henry Ford for the ease of use that revolutionized the automobile industry, just as Macintosh did in respect to the PC in the '80s.[18] Jobs also used the analogy between the automobile and the computer to argue how the capacity for emancipation and liberation introduced by Apple were similar to those found in the spread of the car, that, in his opinion, allowed for freedom of movement. Jobs thus proclaimed the philosophy which would inspire those who want to innovate and understand the impact they can have on society. He adopted the old behavior of the "skilled" programmers in the '70s who possessed unshared knowledge: the user doesn't need to understand the complexity of software to benefit from using an iPhone or a Mac. A few years after Jobs' passing, this is still Cupertino's strategy. It is undeniable that the ergonomic approach of the technological devices was an important factor in making Apple an industry leader,

18 Morozov, 2012, 30.

especially in financial terms. The same goes for the obsession with design and the functional form openly inspired by the Bauhaus school and the style of the German company Braun's products in the '70s. While the technical aspect remains an essential factor, ergonomics and design allow the marketing department to position the firm as the first representative of "luxury" in the technological sector.[19]

Jobs and Apple therefore treat the user as a pure consumer and don't mention the fundamental contribution that common production gave to free or open source software. They never mention the fact that Apple, like most ICT companies, founded their technological proposal on free software and open source platforms. On the other hand, the alleged values of "liberation" bestowed by Apple's innovations and technology deeply contradict the industrial and political strategy of the company. The Californian corporation has always moved in the direction of control through the creation of fences to render user migration to other brands difficult. The choice of creating proprietary devices, operating systems and applications in order to make them inaccessible is a political choice that corresponds to a business model aimed at maximizing profit, just as the direct distribution through the Apple Store has implied the failure of a good portion of its computer resellers. In each new version of the iPhone, the barriers and the incompatibility with market standards increase often proportionally to the lack of any real innovation. Moreover, a business subculture feeds and imposes an atmosphere of mystery and secrecy that gives Apple the vague odor of a sect: we are far from the "Think different" slogan that aimed to oppose IBM's equally sectarian branding: "Think!" Apple understood how to build a particular relation with its users, proposing strong concepts like paradigm shifts or even revolution, tied to its design strategies. One could even ask if there are religious undertones since, with the passing of Steve Jobs, fans flocked to Apple Stores with candles and flowers. The gods of success reside in design and marketing and stores are their churches.

In the case of Microsoft, proprietary logic has traditionally been limited to software. We won't linger over this aspect as much as its

19 An indirect confirmation comes from the recent launch of a Chinese Android smartphone even more expensive than the latest iPhone: it is certainly a top shelf device, but the name of the model just happens to contain the mention: "Porsche Design."

business model that led to a near total monopoly for Windows and the Office package, now in marked decline. As we will discuss later in the case of mobile devices, Microsoft didn't see the tide of the free software paradigm arrive in some technical fields and simply barricaded itself to defend the position it had acquired.

END OF MODERN TIMES

Polling and Interrupt

THE END OF MODERN TIMES BEGINS WITH THE INVENTION OF INTERRUPT. Since its conception, the central unit of a computer has always needed an internal clock in order to function and synchronously cadence the operations of the processor. In fact, computers of the '50s and '60s execute programs one at a time, a functioning called "batch processing": programs are grouped and then sequentially executed. The programmer writes the instructions in a programming language on an external support, for example on punched cards or tape, but does not directly access the mainframe terminals where the program runs. The programmer will then give the program to the calculation center where it is assigned an execution time, a precious resource since every central processing unit (CPU), the equivalent of today's microprocessors, is very expensive. Once the cards have been read, the program is compiled[1] and, if there are no errors, the binary code resulting from the compilation is put into line to await execution. When

1 Compilation is effectuated by a special program called a compiler. A compiler takes in a program, the source code written by the programmer, on which it executes a series of operations as to obtain an error-free object code that is executable by the computer processor.

execution begins, it cannot be interrupted by any external source, because no interaction that could possibly "distract" the computer's CPU from what we could call an "autistic" process is allowed. In the end, if the program includes printing an output, the programmer receives the sheets of dot matrix paper to verify its correct functioning.

However, programs don't always function in a linear and foreseeable way, just like in material reality, constituted of extraordinary phenomena and exceptions. Exceptional states in programs are, for example, impossible operations, like an attempt to divide a number by zero or other more technical situations like an error in the memory allocation. In order to maintain coherent functioning, it is necessary to know how to treat these exceptional states through a special mechanism that allows the computer's central processor to be coordinated with an unexpected internal event, like a programming error, without the computer crashing.

External events are also unpredictable and thus defined as asynchronous, so they too must be somehow managed. To do so properly, the processor must save the state and the context of the "batch" program that it is currently executing in order to deal with these external situations and, once the situation is resolved, must correctly resume the interrupted program. The first interrupt was used in 1956 on a Univac 1103 computer installed in one of NASA's aerodynamics laboratories. The batch functioning of certain programs required instantaneous interruptions in order to integrate data coming from the wind tunnels[2] as they were generated. Through these new interrupt mechanisms, both logical and physical, the processor thus acquired the capacity to manage "stimuli" from the outside world and, at the same time, to answer more and more promptly. This exchange with the outside required the computer's internal clock to be synchronized with a human convention of time, like our Coordinated Universal Time (UTC), or what humans consider real time, from which we derive the definition of real time computing.

From these first steps, the possibility of making computers communicate through multiple channels and devices was born. And while it may seem evident that when we write on a keyboard plugged into a computer the characters appear on the screen, in reality a long process was needed to get to this seemingly banal point. It was necessary to

---

2 A wind tunnel is an apparatus that is used to study the flow pattern of a fluid (for example air or water) around a body.

integrate both hardware and software design mechanisms like IRQ (Interrupt ReQuest) that permit ties between central processing units and an ever-growing number of peripheral devices to be established. In the beginning, they were simple keyboards, displays, magnetic tapes data storages etc., while now we have multiple and intricate connective possibilities with PCs, smartphones and tablets.

With the introduction of interrupt and its evolutions into evermore complex and sophisticated mechanisms, such as interrupt vectors* that are able to manage multiple interactive levels with the outside world, software entered into the "social" era.[3] Programming potential for applications that dialogue with the external resources expanded through a multiplicity of functions and interfaces that put an end to the somewhat solipsistic Turing machine. This expansion took us from NASA's '56 wind tunnel connected by a mainframe to today's "internet of things," a theme that we will return to and that is constituted by a multitude of singularities: human or robotic, active or passive, animated or inert, they are all a part of the network. With the development of the net, these interactions take place in spatial dimensions that range from the nanoparticle inside the human body to probes heading towards Mars.

From an economic and political point of view, the creation of computers that function in real time corresponds to a further phase of development in industrial capitalism. Once it is possible to make programs dialogue with the outside world, why not use them to increase industrial productivity? Real time computers were destined to animate the new intelligent machines of Toyota plants; machines that allow for more minute and more flexible control and exploitation compared to the old assembly line mechanisms of rigid mass production. In a second phase, real time computing will constitute the basis for the development of automated factories.

"In the beginning was the Command Line..."[4]

After "batch" computing with punched cards and mainframes that could not engage in any direct dialogue, thanks to interrupt, it was

3 See the article "Interrupt" in Software Studies (Yuill, 2008, 162) for a complete description.

4 In the Beginning was the Command Line is also the title of a technical article by Stephenson (Stephenson, 1999).

possible to connect peripheral devices to computers. In particular, direct human-machine interaction is established by connecting a terminal that allows for writing and printing. To this end teletype (or teleprinter or TTY) was adopted. TTY existed before WWII and was the terminal used for communicating across telegraph and telex networks.[5] This device was adjusted to work with the first mini-computers near the end of the '70s. Once connected to a computer, which in itself implies the activation of a specific management program on the terminal (today called a "driver"), the TTY was made to operate as in a Command Line Interface (CLI). The operator types a line with a command and, using the "return" button, that line is sent to the computer which can then respond. This type of interaction worked for decades and endured even after the appearance of video terminals. The first PCs still functioned through this kind of textual modality, just like all the first versions of office automation applications.

In 1981, in the Xerox PARC laboratories, another type of interface able to exploit, among other things, the graphic potential of displays was invented. The principle of the Graphical User Interface (GUI) that allows users to interact with the machine by manipulating conventional graphic objects was created. Xerox's invention was commercially deployed by Apple, skillful in capturing other people's innovative ideas, with the almost experimental computer Lisa, subsequently to be launched on a massive scale with Macintosh. Later, Microsoft will commercialize its first version of Windows, while on Unix platforms the graphic interface X Open would be developed.

Starting from that moment, the command line interface became secondary, used for the most part only by programmers and hackers. It still survives, for example, in an accessory program in Windows that opens a command line in MS-DOS, the operating system that preceded Windows. It is also still present in the "terminal" function of Linux and Unix systems where, among other things, a command language (Shell) able to give orders to the system is available. The graphical era broke the heavy chains of rationality imposed by writing and pushed the human-computer relationship to a new dimension. Here we are facing a giant leap forward, even if at first this wasn't clear. A GUI provides a new degree of freedom to users, allowing them an

5 Telex was a telecommunication system developed since the '30s and widely used in the 20th century for commercial correspondence between companies.

exchange that is no longer codified but rather directly sensorial. The mouse, that partially liberated them from the use of the classic keyboard, participated in this shift thanks to an essentially intuitive use, even for children or for people who do not know any command line languages.

Even before the affirmation of video games, this interactive graphical capacity was immediately put to use for productivity in the most widespread office automation applications and "Computer-Aided Technologies" like CAD and CAM, which revolutionized technical and industrial design. The GUI became an integral part of the passage to automation, spreading like wildfire from the factory to the office, accentuating the sociopolitical consequences of the PC's incursion into cognitive labor. It would be reductive to think that the changes induced by the GUI were limited to these aspects. In fact, it became the fundamental tool for cultural industries and creative production, in the widespread, artistic sense of the term. In this field, GUIs have a modality of use that reduces, if not totally destroys, the separation between the creator and the user. This is a tendency that emerges in many sectors of digital technologies, with the fusion of the producer and the consumer in the term "prosumer." GUIs have also become the principle characteristic of the most incisive cultural phenomenon of our era, video games, a subject that will be addressed in depth in the second part of this book.

The recovery of the client-server model

In the '80s, new intuitive graphical tools, both autonomous and powerful, became available on PCs and their cost began to fall, permitting a wider distribution into the cognitive labor of the service industry; an "anarchic" expansion, thanks to the creative possibilities of graphic interfaces and the autonomy of any single worker at the helm of their first cognitive tool.

At first, willingly or unwillingly, management left a margin of liberty to facilitate innovation and change. Later, facing such an ample potential for autonomy, the higher-ups – having gathered all the fruits of this extensive transformation and innovation – searched for a technological key able to reestablish hierarchical discipline and profit maximization. The objective was to rebuild, using the information technology, the hierarchical structure that had been put into question with the arrival of the first PC and later the freedom provided by the GUI. It is here that the model of network application architecture known as client-server was introduced. In this architecture, a vertical hierarchy is restored where "clients" are applications that run on PCs and are no longer autonomous, but depend on one or more servers in a subordinate relationship. The graphic and autonomous environment of personal information technology is exploited by centralized software applications that reestablish hierarchies and privileges, beyond limiting the rights and the visibility of every single user or groups of users, taking away part of the freedom to invent and innovate in their own labor that PCs had given to them.

The administrator of client-server systems has absolute power and great control over the users who are thus subjected to the logic of efficiency and productivity. Dmytri Kleiner, in his "TeleKommunist Manifesto"[6] (Kleiner, 2010), states that networked computation and economic systems can be described in the same terms as social relations in society. The computer networks of capitalism are based on centralized client-server applications to reproduce the principles of privilege and control. Kleiner juxtaposes this architecture to the one based on peer-to-peer (also written as peer2peer or, as we will use, P2P*) that better corresponds to a society with egalitarian tendencies. These hypotheses remain controversial because, beyond their broad, possibly superficial character, technological architectures can be bent to many different uses and, in themselves, never determine any one political organization. In particular, as we will see in the final section that specifically addresses the uses and potential of P2P technologies as organizational strategies, it is important to stress that centralized applications that do not necessary belong to capitalist economic rationality do indeed exist.

High-Frequency Trading (HFT)

A demonstration of how network technology can be molded to financial logic and the potential danger that this entails can be seen in High-Frequency Trading (HFT).[7] Although it has already been

6 See http://telekommunisten.net/the-telekommunist-manifesto.

7 For a detailed technical presentation (in French), see the intervention of an HFT expert (Erra, 2015).

widely criticized, HFT remains an automatic way to intervene in financial markets based on time. It uses algorithms and extremely sophisticated hardware and software based on ICT. The brevity of the transactions is key to the system: investment positions are held for minute fractions of a second. The scope of this approach is that of profiting from extremely small margins, even fractions of cents. To transform such minimal margins into significant earnings, HFT must necessarily operate over a great quantity of daily transactions. One of the main parameters is tied to the speed of transmitting these orders, hence the physical placement of the system is a fundamental factor. So fundamental it has created a "product" sold by specialized companies, including the professional branches of large telecom operators. The flagship product, sold at a pretty penny, is the "co-location," i.e. the possibility to install an HFT system next to the trading targets, for example near the world's largest stock markets. Clearly, the buying and selling of an action in these conditions can be effectuated in terms of milli- or even microseconds. HFT now dominates North American markets, accounting for some 75% of transactions made in 2012, and it is growing in other markets. Due to HFT, the average duration for holding stock has fallen from more than one year in 1960 to 22 minutes in 2012.

The Flash Crash was a sudden fall of about 1000 points, i.e. 9.2% of the Dow Jones, in the space of 10 minutes – between 14:42 and 14:52 local time on May 6th, 2010.[8] The cause was a sudden bug in the HFT systems in the context of a downward trend and sparked by an anomalous automatic transaction. The various investigations that followed highlighted the structural dangers of HFT. With the introduction of machines and algorithms whose aim is rent accumulation in timeframes beyond the capacity of human intervention, an uncontrollable variable that is potentially capable of bringing down the entire financial system in just a few minutes comes into play. The power of this variable would have made even Lenin jealous!

However, the official investigation made by Wall Street authorities curiously fails to mention the initial transaction that started the avalanche, and declares that HFT is a positive modality for trading. The 2010 Flash Crash is not the only episode that testifies to the dangers

8 Flash Crash: sudden collapse of the stock market index following algorithmic trading operations that crash.

of this system because, in August 2012, the Knight Capital trading company lost $440 million in 45 minutes and plunged 140 New York Stock Exchange titles into chaos, an event caused by a so-called dysfunction in one of their algorithms. In this peculiar context, it is easy to fall into the paradoxical: in Paris, a one-hundredth-of-a-second HFT transaction, disputed before the stock exchange authority, required the writing of a 15-page report and a two-hour meeting for an agreement to be reached.

Back in 1996, Paul Virilio, the thinker of speed, had already intuited both the relation between speed and economics as well as – with extraordinary foresight – the type of risk that a reckless use of technology can arise:

> The question of speed is central and is part of the question of economics. Speed is at the same time a threat, as it is capitalized and tyrannical, but it is also life itself. [...] We progress in a new technology just recognizing its specific accident, its specific negativity [...]. Today, it is a general incident, an accident that immediately affects the whole world.[...] But the Internet incident, or that of other technologies of the same nature, represents the emergence of a total accident, not to say integral. And this situation is unprecedented. We have never yet known, apart from, perhaps, the crack of the bags, what could be an accident that would affect the whole world at the same time. (Virilio, 1999)

HFT is a good example of both the liquefaction of space in time, a dimension where the time of human interaction or relations are completely out of play, as well as of the paradigm shift of speed. We move from “time is money” to “acceleration is money,” entering into the sphere of the absolute speed of electromagnetic waves. Consequently, we also risk losing control to a digital Leviathan capable of bringing down the entire system.

The accelerationist movement (Williams & Srnicek, 2015) is inspired by these and other considerations on the existing dynamics and mechanisms between production, technology and life. Accelerationists’ main consideration is the following: since we cannot slow down the speed at which technology effects humanity, transforming it in a less and less controllable way, we should not only accept this increase in speed but even push it to paroxysm, grafting and accompanying it with activities that can help avoid an implosion and favor the emergence of a post-capitalist system. Thus, accelerationists propose we prepare and organize ourselves, counting on the acceleration of technology and not working against it.

One can doubt this hazardous bet: acceleration is the primary variable used by capitalism in the biopolitics of prescribing subjectivities. It is risky and contradictory to fight the enemy with the same weapons and hope of coming out on top. The same risk, but in the opposite direction, can be found in the most fundamentalist theories on degrowth. In both cases, these theories are based on an apocalyptic vision in which human presence on the earth is doomed with or without technology.

From Fordism to Toyotism: real time computing

Let us go back to the consequences of real time computing on the modes of industrial production. Unix and real time systems are the main tools of this passage toward Toyotism and factory automation. As we have seen, from their beginnings, they have allowed for the development of programs and application procedures that manage machines and industrial processes. Starting in the '60s and '70s, real time computing progressively became the keystone in the functioning of industrial tools around the world: machinery, nuclear power plants, transportation and, more generally, any important industrial process. This was a little-advertised but far-reaching revolution: connecting computers to machines became the distinctive sign of a third industrial revolution.

If, on one hand, the 360-system launched by IBM in 1964 was the first widely distributed computer used for automating a part of the cognitive labor at that time, on the other hand new real time computers were perfect for integrating the living labor of the mass worker into robots and automatisms. This not only increased productivity, but also pushed out the young proletariats who were coming into the factories and disrupting the institutional blocks of both unions and management. At the same time, automation also represented an answer to the workers' struggles against alienation and the physical and mental harm that the assembly line caused, for example in the painting and welding departments of large automobile factories. Information had already been incorporated in the weaving looms[9] of textile plants during the first industrial revolution and would then also be integrated into the complex assembly lines. The time was ripe to open up the black box and make computers dialogue not only with humans but also with the assembly line, adding the functions of capture, counting and measure that were previously encapsulated in physical tools or even manual techniques. Even single devices were interfaced with computers enabled with operating systems and applications that were "sensitive" to the outside world. The "interrupt" mechanisms allowed central units to measure unforeseen (asynchronous) information from the outside and then go back to their normal functioning without being disturbed; and now the factory, like the science laboratory, could effectuate operations in a time that was no longer measured on a human scale. In a single second, hundreds or thousands of interruptions allow every single computer to acquire and send information through "industrial" algorithms.

From an economic and political point of view, the role of these devices is that of controlling the neuralgic nodes of production from which the valorized essence of labor is captured and the causes of worker struggle in the factory are neutralized. Real time computing animates robots, thus becoming the key element in emptying the Fordist factory through the automation of the assembly line and industrial production in the largest factories (petrochemical and others), increasing the production of nuclear energy, revolutionizing transportation (high-speed trains, airplanes, supertankers etc.), managing evermore complex and dense networks, all the way to making spectacular space flights. Real time programs are becoming increasingly sophisticated and risky in order to drive high-performance and highly complex systems with a small workforce. Furthermore, real time accelerates the end of the phase of labor exploitation in terms of mechanical energy to expropriate the information produced by workers in the name of modernity and the reduction of costs of making great quantities of standardized products. The driving force of this movement is, as usual, profit, and if new systems are not robust enough and have anomalies that can create small malfunctions or large

9 The Jacquard loom, introduced in 1801, is a type of weaving loom that has the ability to perform complex designs.

catastrophes, it doesn't really matter because the price will be paid by the "collateral victims" of industrial territories: Chernobyl, Bhopal, Seveso, Fukushima, etc....

After the steam era and the introduction of electricity later, real time opens the third and last phase of the industrial era: thermodynamic forces come into play in the first, electricity is the immaterial fluid that irrigates the factory in the second and the flow of information and networks are the nervous system that animate the automated, sophisticated and complex actions in the third phase. Another consequence is the disappearance of "timing and methods" offices, i.e. those caricatures found in films or comedy sketches where obtuse supervisors in white jackets acting as the owner's watchdogs, measuring the ergonomics of every single worker's gesture with a stopwatch in hand with the excuse of "rationalizing" articulated processes.[10] With real time computing, these characters completely lose their function. The information contained in the skilled blue collar's know-how is expropriated and integrated into a robotized machine, as do the repetitive gestures of the assembly line.

If we still laugh at Charlie Chaplin as the guinea pig of the automatic machine in *Modern Times* (Chaplin, 1936), it is precisely due to the absurdity of a robot that instead of alleviating the repetitive alienation of the assembly line, it overturns the entire situation: the machine clumsily takes control and humans become the object, provoking the paradoxical scenes that we know all too well. Thanks to the application of robotized real time computing in the Toyotist factory, these scenes become reality: "it is no longer man who uses the assembly line, but the assembly line that uses man, and we are sure to lose this battle" affirms Satoshi Kamata (Kamata, 1982) following his experience in a Toyota factory in the '70s. Moreover, in sectors where human work remains essential for the time being, such as logistics, the controller in a white coat, hand-held chronometer is replaced by technological and algorithmic devices integrated into the living. Amazon's intentions as the world's leading logistics company are quite revealing:

> At the end of October 2017 Amazon filed two patents for a wristband that tracks the performance of tasks

10 In this regard, see the comic scene of the supervisor played by G. M. Volonté in the cult film of an era *The working class goes to heaven* (Petri, 1971).

> assigned to a worker through an ultrasonic hand movement detection system. The bracelet periodically emits sound and vibration pulses on the wearer's skin. Once an order has been received, the workers must rush to pick up the product on the shelves, pack it, etc. It seems that Amazon considered the hypothesis of the use of bracelets also outside and on cargo ships. The objective is not only to track parcel shipments in real time, as is already done, but also to anticipate and control behavior in a system where humans are governed like robots.... It is for this reason that we speak of "Amabot," i.e. humans considered as "Amazonian robots."[11]

The mirage of Toyotism: lean production

Let's open a parenthesis around a few theories elaborated around *commons-based peer production*, of which the most prominent example is probably the free software movement. The definition comes from Yochai Benkler, the theorist who, inspired by these movements, claims that they are at the base of a new collaborative capitalism, as opposed to the "egotistical" capitalism that currently dominates our lives. Benkler highlights how inside the current system, which he himself does not question, indignant critics predict dire consequences for the future of the biosphere and humanity. These consequences are due to the profound deterioration caused by decades of capitalist practice that is incarnated in the neoliberal organization of cognitive labor. He then correctly observes that, immersed in an egotistical placenta of unbridled competition and cynicism induced by this system, whole generations are conditioned by a pessimist and negative vision of future human relations.

Benkler bases his theory on the possibility of a humanist capitalism, going back to Toyotism, lauding it in one of his best-known essays, *The Penguin and the Leviathan* (Benkler, 2011). Toyota's management emerges with their heads high with respect to General Motors because they earn two or three times less, led by a president who settles for only $1 million p.a., while his American equivalent takes home 3 or 4

11 Roberto Ciccarelli, "Brevetti Amazon, il padrone ti spia," *Il Manifesto* - 02.02.2018. Traduction de l'A. Voir aussi (Malet, 2013)

million. Benkler then discusses the organization of labor, attempting to demonstrate that Toyota's success is due to its ability to cooperate which the multinational corporation establishes within its practices. The ideal cooperative factory is the NUMMI, an old, once declining GM factory that was saved through a joint venture with Toyota taking up the reins in the '80s. Benkler gives a simplified *kaizen*[12] course in which he draws an idyllic and linear picture of the factory's transformation managed by Toyota to illustrate his theory: whereas GM managed the workers' every single movement with a vertical and rigid hierarchy, the Japanese created "collaborative" teams with rotating responsibilities, introducing training courses and problem solving circles where, in theory, everyone has a voice.

He doesn't mention, however, that the break from Taylorism, Toyota's mutations and the robotizing introduced by real time computing are the effects of political decisions. Here we will not take to analyzing Toyota since this has been amply done in the past, sometimes even by heterodox economists who have not been led astray by certain cultural aspects and with a tendency to *penser à l'envers* (Coriat, 1991), losing sight of the purely economic objectives of these operations. We don't know if the nuclear reactors of Fukushima were also "thought of in reverse" but it is certain that today millions of people are subjected to dangerous doses of radioactivity that will impact people's health for generations. Thirty years after the scandal caused by the revelations in Kamata's book on Toyota's factories, we can now evaluate the consequences of "lean production" and the "just in time" strategy.

The miracle of Japan's postwar industry was born from the particular context dominated by the triple alliance of the Japanese state, the American occupying forces, and capital. During the Korean War (1950-1953), the working conditions of the Japanese workers were among the hardest and the quality of life was ranked as one of the lowest in the world.[13] The three-headed powers, worried about com-

12 The term *kaizen* is of Japanese origin and is commonly translated as "continuous improvement." It refers to an improvement path that proceeds with uninterrupted constancy and that involves the entire company structure. The term *kaizen*, in fact, is the composition of two Japanese words: Kai (renewal) and Zen (way).

13 "In Japan, the working day in industry is one of the longest in the world and the quality of life is low and does not improve. Public criticism increases in

munist influence, destroyed the unions structured by category and imposed engineer Ohno's productive model. This is the model at the root of Toyota[14] and would fulfill the needs of the North American automobile market. Within the chorus of those who denounce "lean production," we find those who explain the blunders with the fact that the fundamental distinction that pervades Japanese business culture was not understood: *tatemae* (what you are supposed to feel or do, a kind of superego) and *honne* (the ego, or what you actually feel or do). Here is a 2006 testimony from Darius Mehri, an American engineer specialized in computer simulations and who worked for three years at Toyota:

> What has changed at Toyota over the last three decades? Not much. I experienced the same unsafe work environment, the same oppressive mechanisms of worker control, the same power manipulations that Kamata chronicled. He described the same *tatemae/honne* disconnect that was pervasive in my experience. [...] There are some who continue to accept the *tatemae* without understanding the *honne* it belies. Toyota was recently lauded for the reduced design time in producing the Prius. But never is the impact on the health and safety of engineers mentioned. Like I was, I suspect engineers and production employees were simply pressed, intimidated, and overloaded to get the job done.[15]

Benkler maintains that the principles of lean management born in Toyota must today be applied to all working processes in biocognitive capitalism. It is no accident that he joins this praise of Toyota

---

comparison with the 'lean system,' with regard to the waste of human and natural resources and to a system that increases stress and makes life congested and polluted places of life. The industrial organization is conceptualized as a profit-making machine that depends on the effectiveness and satisfaction of the customer's needs." Stuart D. Green, *The Dark Side of Lean Construction: Exploitation and Ideology*, 1999, 24.

14 See also C. Marazzi, *Il posto dei calzini. La svolta linguistica dell'economia e i suoi effetti nella politica*, 1994.

15 (Mehri, 2005, 41).

with the generation of open source coding, comparing the passage from Fordism to Toyota's lean production with that from proprietary software to open source. Numerous examples corroborate his reasoning, like when he cites the transition from Encyclopedia Britannica to Wikipedia, or when he commends Google (Benkler, 2011, 10).

The lean management movement in the ICT sector comes straight from Silicon Valley where, beyond Google, old and new corporations use technology to test these new, lucrative models. This is the idea that Benkler tries to prove, affirming that a new capitalism is possible thanks to renewed cooperative forms that should triumph over egoism without saying a word about the financial structures of digital multinational corporations. In this hypothesis, the understanding that industrial capitalism is based on exploiting forms of cooperation in the factory is forgotten. Whether it is Toyotism or the businesses in Silicon Valley, cooperation per se doesn't change the foundations of political and social relations.

The principles of lean management have already been amply transposed and adapted to cognitive production in general and, more specifically, in the working conditions of digital labor. It is precisely this sector that Benkler uses to theorize how a commons-based peer production operating inside biocognitive capitalism should save us from the Leviathan of the hierarchic capitalism that, according to him, is taking us to the brink of disaster. While we can undoubtedly adhere to this last observation, we find the first part of Benkler's affirmation misguided: skyscrapers in business districts and workers' computers are full of so-called "collaborative" software. The "ethics" charters of digital corporations mention, exalt and, at the same sime, attempt to bridle the naturally collaborative spirit of project teams. Despite appearances, suffering, precarity and psychopathologies are on the rise, precisely due to a compulsive and obsessive "just in time" strategy that feeds capitalist competition.

Information technologies in cognitive labor control and management

The transformations introduced by Toyotism cannot be considered technological innovations that fell from heaven from a generic idea of "progress." Two decades of worker/mass-consumer struggles in the '70s-'80s put the Fordist model, characterized by sheer volume, rigidity and standardization, into crisis. Ultimately, they pushed capital to introduce automation and robotics in the factory. A new historical era of an economy founded on knowledge was opened. In the transition between the industrial and cognitive phases, the new figure of the social worker developed, where the relation with labor, characterized by a new mobility, radically changed. Andrea Fumagalli describes this mobility:

> The mobile condition of the workforce is accompanied by the assumption of individual bargaining. This is due to the fact that nomadic individuals are being put to work and the primacy of private law over a common law still to be established induces the transformation of individuality, especially if characterized by cognitive, relational and affective activities in individual contractual terms. It follows that the intrinsic mobility of work is transformed into subjective precariousness of work (Fumagalli, 2015).

The concrete relation between multitude and singularity, originally Spinozian, is used in Negri and Hardt's quadrilogy[16] to represent the new complexity of the composition of the workforce, no longer reducible to an unqualified mass like in the industrial era. In the same way, it is harder and harder to liquidate the homogeneous stock of impersonal merchandise from assembly lines in this context.

The emergence of the personal computer in the '80s helped to break the mold of industrial innovation and renewed the myth of garage invention.[17] In particular, Apple revealed itself under a different light in respect to Big Blue (IBM), positioning itself as a kind of precursor to today's anarcho-capitalism. This image was further developed in their advertising. The most famous example is the Orwellian launch of Macintosh in 1984 or the urban legend, cunningly kept alive, of a logo inspired by the cyanide-laced apple used by Alan Turing[18] to

16 (Hardt & Negri, Commonwealth, 2010), (Hardt & Negri, Multitude, 2004), (Hardt & Negri, Empire, 2001), (Hardt & Negri, Assembly, 2018)

17 A myth born with B. Hewlett and D. Packard who started their business in a garage in 1953 in Palo Alto (CA).

18 Alan Turing poisoned himself in 1954 with a cyanide apple following a conviction

kill himself following his conviction for homosexuality. The fact is that PCs were the first machines that represented the affirmation of the mass intellect. From the mid-80s onwards, their production was industrialized and expanded to the point of making them accessible to large parts of society. Even though PCs still weren't connected or organized in local or remote networks, three axes of influence emerged: code generation, office automation and gaming.

For the first time, a personal machine infinitely more powerful and flexible than the typewriter is made available. In the early '80s, after an initial shock, the interest and enthusiasm of a new working generation discovering the incredible potential of these new devices is mind-blowing. The power of a computer is no longer limited to a handful of specialized experts: young laborers, students, teachers, secretaries and service industry workers are astounded by the office automation functions like word processing and other simple applications that are today taken for granted. This general techno-scientific knowledge creates the ideal conditions for autonomously harnessing this new potential. The requirement for knowledge often passes above and beyond the company itself, thus anticipating today's Bring Your Own Device (BYOD) ethos (that will be addressed later), where cognitive and creative workers have to provide their own tools. Tangible proof that the PC is a tool for independence for social cooperation in the workplace and the booming service sector.

Management cautiously adopted the PC for productivity but didn't love the spaces of autonomous organization where its role as Grand Organizer is questioned. Ever since digital applications were introduced in the era of mainframes, they were rigidly dedicated to specific functions like accounting or industrial production management, while other mid-size businesses often relied on external computer services. Unlike the PC, mainframe functions were rigidly centralized. The personnel engaged in their use had specific terminals that were used exclusively for these purposes. PCs overturned this way of working through office automation and word processing, eliminating the typewriter and then, with electronic spreadsheets (like Excel), slide presentations (like PowerPoint) and the first personal systems for database management, they created a context of autonomous

for homosexuality and the chemical castration to which he was subjected. See the film recently dedicated to his story, "Imitation Game" (Tyldum, 2014).

productivity that was no longer rigidly tied to company programs. This uncontrolled creation of value was looked down on by management, which tried to retain their strategic position by reinforcing internal control.[19] The introduction and spread of local networks (LAN) and the Client-Server architecture, allowed for the reestablishment of a new form of centralization that integrated the characteristics of graphic interfaces (Macintosh, Windows, Unix, etc.). The steady introduction, beginning in the '90s, of Enterprise Resource Planning packages (ERP) became the neuralgic center of this operation to take back company control, be it in the public or private sectors.

ERPs are, in fact, global application systems that integrate all internal and external information that moves through the whole of the company, indispensable data for the new forms of generalized control. In addition, ERPs constitute the machine that shareholders demand for optimizing rent through the financialization of capillary companies. Articulated in all sectors of administration, from guiding production processes and personnel management to client relations, it saps vital energy from the cooperation of living labor in office environments. In this sense, a parallel can be established between ERPs in business management and the introduction of real time computing in the factory. Following these two transitions, management can integrate these two tools for control and capture value from these two activities in a more and more efficient way. It is through this immaterial dispositive that the *valorizing* information[20] studied by Romano Alquati in the Olivetti factory in Ivrea is bridled: "productive labor is defined in the quality of the information elaborated and transmitted by the worker to the means of production, with the mediation of fixed capital" (Alquati, 1961). In other words, operating as a digital interface between the domains of knowledge and capital, software code transforms raw information into value.

Again paraphrasing Alquati, ERPs globally and organically recompose all the functions of material, informational and cognitive workers to then transfer them into a Planning. On one hand, ERPs amass workers' existing shared know-how: this is particularly visible in the phase where the software itself is designed and when procedure

19 Carr, *The Dark Side of the Web*, 2008, 193.

20 See M. Pasquinelli, *Capitalismo macchinico e plusvalore di rete: note sull'economia politica della macchina di Turing*, http://www.uninomade.org/capitalismo-macchinico.

"experts" extract functional protocols from workers' knowledge. The experts then transcribe specific documents that are given to the programming team so that they can be translated into an application. On the other hand, ERPs tangibly implement management's answer to the precedent labor struggles. This new system confirms the definitive decline of the technical division of labor between planning and the task's fragmented execution theorized by Adam Smith two centuries before.

But the ERP is, per se, an inert and impotent informational machine and if it is not properly fed, it doesn't function. The old axiom of first American programmers of the Beat Generation is still valid: garbage in/garbage out. In big projects and large companies the installation and execution of an ERP's automated immaterial processes of operative control are the keystone in the transition to cognitive production. The initial adaptation of the application package can be a delicate, risky and expensive operation. The basic version consigned by the software publisher is dead labor, as an unfinished and incomplete assembly line. The true energy that makes it work is not so much the electricity it needs but the information. In big projects and large companies, in order to make the ERP package operational, it must be precisely assembled by parameterizing it, piece by piece, data by data, adding the gears of the specific and complementary developments that are missing. Then, as is the case for all complex and articulated systems, it is necessary to test it for the integration of individual components and their whole to verify their good general functioning and robustness in regular use over the medium term. This process is often called customization of an ERP package and the procedures rendering it "productive" must pass coded tests so that there are no longer any critical or serious anomalies. In France, to use "industrial" terminology, the proof of quality is delivered in a VABF protocol (a term adopted from the French *vérification d'aptitude au bon fonctionnement*). The final acceptance may take a few months, until a VSR protocol (*vérification de service régulier*) is signed.

Interfaces are also worth mentioning since ERP modules and major functionalities (e.g. sales or billing) generally are not always introduced simultaneously. In this way, the company information system can be seen as kind of puzzle in which units are added or replaced. Each module is usually fed and, in turn, feeds the contiguous ones. For example, a billing module will be powered by the sales one that registers orders which, in turn, will feed the accounting and the customer relationship management (CRM) subsystems. The input/output interfaces are one of the most difficult aspects to be implemented and tuned. First of all, assembly is a delicate exercise that must find a compromise between customizing the standard package in order to implement the existing processes of the enterprise and/or adapting the processes encoded in the package to the company; in other words, a balance between the capturing of knowledge concerning internal processes, workers, and those that the software publisher provides. These are usually procedure extracts of its customers already acquired in the form of generic mediation.

This customisation is not a neutral operation: depending on the characteristics of the company, its cultural profile, its (geographical) extension, its business sector, its complexity, the project can last months or even years and involve extremely high costs, which often form a consistent part of fixed capital (furniture, hardware, etc.). It is however important to stress that the ERP machine requires, even more so than assembly lines, not only sophisticated maintenance, but also and above all a continuous stream of tuning, modifications and additions. This is needed to adhere to the constantly evolving processes which conform to the dynamics of business and innovation such as the launch of new products, legislation, etc. These operations are an enormous source of revenue for IT services companies operating in this field. Even today, after decades of experience, there are always risks to the success of projects of this kind. If, in the industrial phase, mechanical machines crystallized a number of operations extracted from *know-how* of the craftsman or worker, the introduction of an ERP amplifies this operation: it is the whole complexity of the processes of the company that is integrated into a single IT environment. Unlike mechanical machines (which reduce manual labor to a series of fragmented and/or repetitive operations), software applications force an interaction with the entreprise information system that is constrictive, articulated, stimulating and irritating at the same time. In fact, it contains a form of subsumption of living labor. The other big difference concerns the way the change is introduced: the industrial component is inserted in the factory and set in a prefabricated static block, while the ERP machine is built on site and with the active participation of those who will have to use it. In fact, it cannot be implemented without the contribution of pre-existing knowledge, without the acquisition of procedures, without opportunely placating the contradictions and conflicts of power. An entire "business services" sector has been created to extract workers' knowledge, translate it into data and appropriate code and to cover other indispensable tasks such as corrective and progressive maintenance, etc. Such consulting and IT services employ millions of employees worldwide, with impressive growth in emerging countries, among which India stands out. Often these operations introducing or reengineering systems generate resistance from target workers because they feel a loss of freedom and theft of value and knowledge. This makes it difficult for the functional consultant who intervenes to try to impose management's business process on the target cognitive workers.[21] This imposition often clashes with the workers' practice of autonomous cooperation using homemade software tools, personally tailored thanks to the pliability of PCs: spreadsheets, specific programs and the use of network applications. These custom application packages generate new short-circuits in productivity and push workers to become either guinea pigs or scapegoats, subordinate or accomplice.

These changes in production are accompanied by the evolution of control and subjugation of cognitive producers through a new hierarchy and new methods of command that attempt to enter the worker's "intimate" sphere, generating widespread pathogenic effects: Didier Lombard, the former Chairman and Chief Executive Officer[22] of France Telecom, defined it, with unconscious cynicism, as "trendy suicide," only later to be incriminated by the French justice system for harassment.

When encountering strong resistance, company executives appeal to the intervention of "change management" consultants, experts in forcing workers to swallow policy changes that are often alienating and aimed at regimenting cognitive production. The numerous cases of partial or total failure of these business planning "transformation" projects are often due to the accumulation of negative factors involved in the transition period; the ERP is not the cause but rather the catalyst and the accelerator. The ERP plans, integrates and subjugates the mechanics of automated fabrications and

21 For example: the secretary, the storekeeper, the maintenance technician, the accounting assistant, the call center employee, the nurse, and so on.

22 In French, *Président Directeur Général*.

cognitive processes towards the maximization of income and profit in the company ecosystem. Cognitive neuroprocesses, more so than the repetitive physical exercises of the assembly line, are now being shaped through the interaction in an all-encompassing technological ecosystem, punctuated by the continuous scanning of databases, by the executive flow of the machine code and the pervasiveness of the access points to the network in space-time. The implementation of these systems, called upon by program/project directors and CIOs (Chief of Information Office), launches a valuable engine that functions thanks only to the living knowledge of company workers. These corporations now have a nervous system programmed to perform orders of a brain focused on financial performance. Under these obsessive conditions, the nervous system cannot be balanced: on the one hand, control functions are so invasive they contribute to such suffering that they sterilize living labor; on the other hand, the system shows signs of schizophrenia when it works to the almost exclusive benefit of management.

The machine that innervates financialization inside the company-body, progressively invades the State Welfare Organization. National or local agencies make use of these internal transformations as a key to privatization. Despite difficulties and resistance, the first phase of the information transition is completed towards the end of the last century. The dominant organization of industrial capitalism dies and from the toxic embers of its waste bio-cognitive capitalism rises like a phoenix. However, this does not mean that industrial or even archaic and proto-industrial modes of production, especially in so-called emerging countries, do not survive and do not co-exist. Perhaps management had not completely predicted or calculated it, focused as they were on the Holy Grail of *return on investment* (ROI) but, despite the technological and social barriers, the processes of the ERP do not remain restricted to the company. They are destined to become a bridle for global networks: sensitive data remains protected and circumscribed but companies begin to develop the means to capture the outside information that interests them. Here we are at the beginning of the movement that, with the development of networks, will lead to the phenomenon of "big data."

The firm's core: marketing and CRM23 – the customer centric strategy or the commodification of the social user

With the fall of the Berlin Wall, barriers surrounding business also fall and production spreads throughout the metropolis. The barrier between producer and consumer falls as well. In the same way, the end of the figure of the mass worker also marks the end of the mass consumer. Rigid and impersonal mass production is inadequate for meeting the needs of the new social agent forged from generalized techno-scientific knowledge. Today, there are still artefacts that paradoxically indicate this change: take, for example, the commercial success of cars inspired by the cult models of the "boom" era like the Bug, the Mini or the Cinquecento. This skillful operation, based on a mix of technology, a retro aesthetic and marketing savvy, masks two completely different realities. Today's models are designed for the privileged or young go-getters that can afford them: they are fast, sophisticated, customized to the extreme and very expensive; the originals, a reflection of the new social categories that allowed a greater degree of freedom, were simple, uniform, robust and inexpensive. The Trabant, for example, with its plastic bodywork produced until 1989, twenty years behind the cars produced in the West, is the star of a Berlin museum dedicated to the German Democratic Republic (East Germany). It has become a collector's item because it has become the failure symbol of state capitalism in Soviet countries.

This paradigm shift destabilizes the profitability that industrial capitalism obtained in the thirty years of the post-war period, when industry was focused on creating consumer flows with higher quantity than those of all previous eras. In the context of great volumes made at low cost, the main sales strategy is price. The latter is low because it is based on the rigid automation of the assembly line, on the fragmentation of labor, on a docile workforce with low wages and on large markets. The company able to manufacture for the largest marketplace at a lower cost was usually the one that conquered market leadership. The new paradigm of cognitive production requires a 180° change in perspective. Businesses make use of marketing, a conceptual tool that had emerged in another difficult situation: in the 1929 crisis, competition had increased and it was therefore necessary to find new

23 For marketing as "the soul of the enterprise," see (G. Deleuze, *Post-scriptum sur les sociétés de contrôle*, 1990).

commercial weapons to standout and survive. Marketing must use all "actions that aim to foresee or ascertain and, if necessary, to *renew, stimulate or arouse* the needs of the consumer."[24] Marketing offices become the bridge for management to project influence, conditioning and sometimes controlling the continuous adaption of the productive apparatus for these newly determined goals.

The first qualitative approaches inspired by Freud and Piaget's "*méthode clinique*" date back to this period. This method of establishing analyses and diagnoses is based on telltale signs in the speech of the patient rather than on the declarative rational aspect.[25] Qualitative marketing specialists bend these cognitive diagnostic modalities for commercial ends, using "panels" set up specifically for this purpose. Later, new procedures more suited to a society based on knowledge (directly implicating and involving the client) are put into place. First of all, a progressive ontological transformation of the client is carried out, shifting to the idea of consumer. The term "client" defines a social relationship, from the Latin "*cliens,*" or a free person under the protection of a more powerful benefactor. But today the client is only characterized by a relationship created through commercial transaction and, as a consumer, is defined exclusively by the act of consuming and therefore through the partial or total destruction of a type of good or service.

It is the consumer who is studied in marketing, first in terms of desire and then in the act of consumption. This is where the necessity for data on opinions in order to better influence or willfully manipulate arises. To develop the essential activity of understanding consumer behavior, businesses create new profiles of workers who are often precarious. They function as *workflow* agents who manage cognitive labor and their immaterial tools consist of an amalgam of procedural skills,

---

24 In the definition of marketing in the French *Journal official* of 2 April 1987 "the set of actions that have the objective of predicting or finding or [...] arousing or stimulating [...] the needs of the consumer" [our translation].

25 The application of qualitative interviews began in the United States as early as the '30s, probably facilitated by the economic crisis of that period. In a qualitative survey at Western Electric, noting that the answers to a questionnaire are not very reliable and contain little information, the interviewers, Roethlisberger and Dickson, abandon the method of the questionnaire with direct questions to move on to a more indirect way where, once the subject has been introduced, the interviewee is freer to express himself.

language, culture, affections and emotions. It is quite understandable that in the tension of transforming and regulating an ever-wider spectrum of activities and exchanges in pure business, the system assigns as the central role the figure of humans as consumers. In the construction of a taxonomy of the business, marketing determines two large categories of clients, in jargon defined by the abbreviations B2B, business to business, and B2C, business to consumer.

The first denotes all public or private entities and comprises all types of production, including "man for man" production like healthcare and education. In the second, the role of consumer replaces and incorporates every other human figure: the patient, the tax payer and the user all become customers. This separation allows company strategists to apply specific modes of influence and control to each category, including marketing strategies, algorithms, system architectures and mechanisms to capture social labor.

A second step consists in involving and activating the consumer thanks to network technologies. Through an alleged gratuity of services, the individual becomes – consciously or not – a direct producer of their consumer subjectivity while simultaneously leaving indicative traces of this process along the way: the *prosumer* is born. Biocognitive capitalism is constantly looking for income through information about behavior, emotions and the life of prosumer in general. However, this comes into contradiction with pauperization, precarity and cuts in social expenditures that undermine consumption. The data collected also aids in the strategic objective of securing the reproduction of the system itself. The so-called "consumer centric" business strategy is sold as a cultural change that transforms the "selfish" corporation into an "altruist" and attentive force for precious customers who, according to the marketing-speak, are "kings." The consumer is no longer prey to hunt, but certain elements in this language betray the real intentions: the client-capital must generate rent. Capturing value is the key to the becoming rent of profit, an integral part of new processes of innovation in biocognitive capitalism.

The avant-garde mobile phone industry: from the customer experience to the NSA scandal

In the '90s, when the European GSM standard was perfected, the mobile telephony industry aligned with changes in work and life and became a core ICT sector. The emergence of the new service industry became the testing area and spearheaded a wave of privatization activities that affected the global telecommunications sector. The procedures for creating the new sector included the sale of telecommunication frequencies to the highest bidder. Big name industrial and financial companies looking for opportunities and reconversion, like Olivetti in Italy and Bouygues in France, come into play.

As a result, the new telecom industry established a constellation of national oligarchies in which three or four operators divided up the old fixed telephones and the new "prize" mobile services in each country. This was such a rapid and homogeneous movement on a global level that it reveals, in itself, a clear paradigm shift. The fact that every mobile operator is destined to manage large volumes of users from zero, puts them in an ideal position for concretely experimenting with new marketing theories. An outward opening is called a "multi-channel strategy." This signals the arrival of call-centers, born in the '80s and which are the symbol of the "offshore cognitive factory," based on the technology sold by these operators.

To launch the newly born mobile, operators needed to improve the internal management of the ERP so that the largest number of "consumer" clients could be captured. Network technologies allow for the development of tools suitable for finely and individually managing relationships with large volumes of customers: these are the testing grounds for the nascent CRM (Customer Relationship Management) applications and packages that will define the sector. Commercial, digital and financial management departments, along with marketing and logistics are activated in order to launch a series of processes aimed at the external world and able to acquire incoming data (inbound) and to promote marketing campaigns (outbound). We witness a race to open channels of communication: in addition to websites dedicated to finding new customers or the management of existing ones, the potential of call-centers is increased tenfold with the introduction of the Computer Telephony Integration (CTI), which consists in programmable automata that allow telephone exchanges to be directly integrated with CRM applications. The CTI, widespread at all levels from small businesses to public administration, prefigures ways of identifying, tracking and retrieving information from customers and from the outside world in general. IVR subsystems (Interactive Voice Response), or automaton telephone programs that, through prerecorded voice messages or synthetic voice, force a man/machine dialogue whose primary purpose is to reduce call-center labor costs while filtering unimportant calls in terms of profit. IVRs are also famous for being a favorite technique in numerous gags. The exorbitant number of possible questions and interactions and the randomness in recognizing natural language make this tool often highly-irritating and ineffective, thus inspiring comedians from all over the world.

The mobile telephony sector is the spearhead of CRM software packages for these experiments at the end of the last century. In this regard, an anticipatory anecdote: in the '90s an Israeli company that provides billing software and IT consulting services for US telecom operators was ordered to carry out the tap and search call "tickets"[26] by the judicial authorities and the FBI. The fact that a foreign company could access sensitive information involving state security caused a scandal, a precursor to what happened recently with Edward Snowden's testimony against the cyber-espionage practiced by the NSA[27] with the Prism program. A precursor in the sense that Prism carries out research in the big data of all the great American multinational internet corporations, as well as in the databases of telephone operators.

26 A "ticket" is a data package that contains information about a single call (for example: telephone numbers and user geolocation, call duration etc.).

27 A former Central Intelligence Agency (CIA) agent and technician at Booz Allen Hamilton (NSA consultant technology company) until 10 June 2013, Snowden is known to have publicly revealed details of several mass surveillance programs of the US and British government, hitherto kept secret. Through collaboration with Glenn Greenwald, a journalist from *The Guardian* who published a series of complaints based on his revelations in June 2013, Snowden revealed various information about secreted intelligence programs, including inter-state telephone interception. The United States and European Union on communications metadata: PRISM, XKeyscore, Boundless Informant and Bullrun and the British Government's Tempora Muscular and Optic Nerve programs, attracting considerable attention from the public and the rest of the media. Snowden said: "I do not want to live in a world where everything I do and say is recorded [...] my only goal is to tell the public what is done in its name and what is done against it." *The Guardian* (London), 9/6/2013.

EXPROPRIATION OF NETWORK LABOR

Investigation into large IT services companies

The expansive dynamics of networked labor in large companies must first be investigated before attempting any analysis. After the initial boom in the ICT sector, the modalities of online labor now encompass the whole of production. We have just seen how and why ERP packages were introduced. Later, once the enterprise internal functioning was under algorithmic control, outside projection was implemented through CRM packages that added interactivity features with masses of client-consumers. Only afterwards is the client transformed not only into consumer but also, as we have seen above, into an unpaid producer (prosumer).

Over the last decade, these abilities to influence, capture and control network labor have been improved by integrating the "collaborative" procedures of Web 2.0 and the pervasiveness of the "always connected" devices with ERPs and CRMs. Another aspect concerns the centrality of project management in business organization. In this context, methods and procedures which integrate management with networked devices (designed, developed and tested initially in ICT companies), such as in system integrators and IT services companies and the giants of Web 2.0, have now spread to all other sectors, from media to finance, from industry to commerce.

Consequently, the alleged dichotomy between digital and traditional capitalism falls away with the emergence of a growing number of activities guided by these IT applications and infrastructures, even in those businesses most representative of "old" industries. We have chosen IT services companies, where the totality of labor is networked, as the field for inquiry into digital labor. In order to reconfigure itself best while facing today's political, financial and social dynamics, this complex system of capture and control must also be continuously regulated, hierarchically structured and updated. IT services companies have large financial and technological resources that generalize the normalization of cognitive digital labor. We will examine in detail how this standardization is also accomplished through offshores, cloud computing,* the circumvention of national labor laws, the constraints of open space offices and persuasive marketing methods. We will then take a look at the role of unions, often reduced to being just another cog in the system.

The international division of digital labor

In the IT services sector, delocalization is called *offshore*. Over the past few years, this practice has been refined by large American, European or Asian companies. Although the global number of *offshore* engineers is constantly on the rise (the vast majority of which are Indian), in some non-English-speaking countries, France for example, attempts to persuade managers with important economic incentives to use this workforce as much as possible has not, so far, been very successful. In France, a first attempt to make local engineers collaborate with Indians on the same integration project was abandoned and, paraphrasing the famous Chinese slogan applied to Hong Kong "one country two systems," the "one project – two culture" methodology did not work: what incentives are there for Indian workers, paid only a few hundred dollars a month with very few contractual rights, to reach a European level of training and productivity*? What control mechanism can be effective for measuring the productivity of individuals integrated into a single project yet separated by continents, time zones, languages and cultures?

Learning from these lessons, IT services companies have shifted from hierarchical management to contractual strategies: offshore teams become subcontractors for specific tasks for which they are wholly responsible and that are entrusted to them using ad hoc contracts. For example, within the scope of a large European project, an Indian team is entrusted with the task of "manually" executing the boring and repetitive validation tests of a banking application. The Indian computer worker performs the most basic job with the least "added value," while upstream the more "noble" tasks of concept, development and final integration are entrusted to European teams.

With the organization of emerging countries and thanks to the impulse of the BRICS,[1] things are changing. Large, multinational IT services are now based in India and compete with those of the developed world. Likewise, a new phenomenon is emerging in technological companies: after the invention of the multinational company without factories, initially theorized by the French telecommunications company Alcatel,[2] we have now arrived at the point that they no longer even have their own research centers, with research being completely outsourced or sent offshore. Faced with a decline in vertical and centralized architectures, and the affirmation of free and open source software, companies use the free software as a "positive externality" (i.e. something that costs nothing and from which profit can be drawn). This resource by itself, however, is insufficient. The invention of cloud computing* (which will be further discussed in relation to its role in organizational strategies), is another step in this direction, since it consists in entrusting a private company, called a "Service Provider," with our own applications and data, making them executable from the cloud, i.e. a virtual infrastructure (servers, networks, etc.) that are somewhat opaque and difficult for clients to control. Many denounce this practice as pure marketing strategy but it is, above all, a giant business model which implies the expulsion of labor from IT services in order to accelerate outsourcing practices and the physical elimination of millions of servers and public or private IT centers. The aim is to centralize computing power in to the immense, hidden and anti-ecological server farms of Google, Amazon, Microsoft, etc. In addition to the loss of control of our own information, rightly denounced by Stallman (the founder of the Free Software Foundation), this operation seems to be configured as a kind of informational communism of cognitive capital.

---

1 Acronym for Brazil, Russia, India, China and South Africa.

2 Alcatel-Lucent was acquired by and merged with Nokia in 2016.

Over the short- and medium-term, cloud computing responds to the imperatives of a company's financial department to reduce drastically fixed investments (CapEx) in production systems – of which IT is an essential part – and to dilute them into the more flexible and controllable cost of managing an external, perhaps offshored, service (OpEx).[3] It is another central element that macroscopically highlights the strategic "seizing of power" within the corporate governance of the financial model and, consequently, of organizational instruments and management methods designed to make this model operational.

IT services (or consulting) companies refine their value extraction techniques both from common production as well as from direct exploitation of networked cognitive labor. New parameters now consist in modulating the activity according to each national context, taking into account economics and cultural and legal dynamics, with particular regard to tax and labor laws in every country. On the other hand, the exploitation of local tax laws in their favor in collusion with the executive is another typical mode of operation of large multinational corporations, including GAFA and related industries, which appear in the *Luxleaks* scandal. Less well known are the practices concerning labor law, even though IT services companies are at the forefront in exploiting this legislation to their advantage.

This is clearly explained by the founder of the first European and French IT consulting companies or SSII (Sociétés de Services en Ingénierie Informatique)[4] in an interview: "it is easier to hire an Indian than a Frenchman. First of all, because he speaks English well and does not count his hours (of work). And if one day he realizes that there is no more work for him, then, pragmatically, he tries his luck elsewhere. In France, when we hire someone, it is for life..."[5] What can be gleaned from this declaration is not the intention to abandon Europe, where tens of thousands of wage earners provide substantial profits, but the desire to transform most of these laborers into

3 In finance jargon, "Operational expenditure (OpEx) is the cost necessary to manage a product, a business or a system or operating and management costs. Its counterpart, capital expenditure (CapEx), is the cost of developing or providing hard flights for the product or the system.

4 Acronym of French origin, where this sector is particularly developed, and that we will use in this chapter.

5 P. Bonazza, *Yes on the dort on est mort*, Le Point, 03/06/2010 [our translation].

precarious workers, just like their Indian counterparts. To achieve these objectives, IT services companies adopt an organization capable of adjusting their policies between the local and the global with extreme precision.

In France, for example, where social and legal protection was until recently still relatively consistent, profit levels are maintained by increasing and selling the labor of numerous trainee engineers and apprentices with heavy state subsidies aimed at containing youth unemployment. Another stratagem is dumping the economic responsibility for seniors onto the social collectivity: as soon a senior can no longer guarantee sufficient profitability, they are fired and, despite unemployment welfare, it can be difficult to reach an increasingly advanced retirement age…

In European countries where labor has already been made more flexible, there are other opportunities that arise from the dismantling of the public sector. An IT consulting company for example can win a bid for outsourcing national tax information services. In exchange for ten-year, multi-million Euro contracts, thousands of state tax services employees are privatized and rendered precarious with very few employment guarantees. Later, in the name of efficiency and profit, most of these workers will be fired with greater ease and the services are appropriately relocated to India.

Business marketing and "ethics"

"Working in a contemporary enterprise means belonging – joining its world, its wishes, its beliefs" writes Maurizio Lazzarato.[6] In just a few years, the subjugation dictated by financial governance has so invaded labor, even in the most minute interstices, that the world of IT industry, like those of many companies in the cognitive sector, have become a nightmare.

For young graduates in science or technology who, until just a few years ago, generally went into their first work experience with interest, are now faced with unpleasant surprises. While a lucky few find engaging projects, most find themselves doing tedious or constrictive jobs on a daily basis, like spending their first two years doing repetitive corrections on old programs or having to start at 7 a.m. in order to

6 Lazzarato, 2004, 108 [our translation].

ensure hourly coverage of a customer hot-line. Demotivation is one of the reasons for the explosion in turnover rates. This situation is completely reversed in comparison to the recent past when, despite the paradigm of profit, the necessity to integrate employees into long-term business projects still existed.

Faced with the lack of attractiveness and motivation and such a degraded company image, enlisting trainees and new graduates has become an arduous task, so company management decides to use marketing to make up for unattractive aspects of the work environment. Investments are made in Human Resources (HR) to make them profit centers in recruitment. It is no coincidence that the people in charge of recruiting are often called "business resources partners" and their remuneration is variable and can be linked to the quantity and quality of the "resources" hired.

In the "industrial" age, these offices had an almost exclusive function of control and repression, a kind of Ministry of the Interior for the company. Today, these departments add to these functions communication and data collection through algorithms, technologies and applications, "business intelligence" and data mining in order to perfect candidates' profiles. The latter are now "customers" of which HR wants to know as much as possible: in addition to classic information (age, sex and education), their profiles now include a multitude of important indications like race, health, religion, family life, affections, delusions, hobbies or technological orientations (e.g. football fans as opposed to geeks) and will also probe the notorious criterion of "stress resistance."[7]

This and other models also have a boomerang effect, in the sense that young people know how these mechanisms work and touch up their résumés, taking pictures in specific poses on social networks in order to send appreciable signals that allow them to get a job, somewhat like in previous generations, when hippies or "revolutionary" militants showed up at interviews in a suit and tie, or like when a "normal" couple signed the lease for a home. The main objective of this propaganda is to hide the brutal realities of the business world today from young people and to limit the turnover that, in some IT sectors, has risen from 5% annually to 20% or even 25% over the last two decades. This last point (which will be addressed later when

7 This is an ever more important criterion used in the periodic personnel evaluations.

analyzing the phenomenon of nomadism), seems now acquired both by management and young graduates, and especially those coming from the extra-European countries that are often the target of xenophobic measures adopted by governments of the EU. Aware that they have few career opportunities, if they manage to avoid being expelled after their expensive studies, they limit themselves to one or two years in a big company in order to acquire a minimum of experience and to add a "recognized brand" to their résumé.

Hiring campaigns are conducted in refined ways and dispose of great means, including salons where, in enormous spaces, IT services companies try to show off their alleged attractiveness. Communications also come across to social media networks, with slogans like "My work, my life,"[8] which irresistibly evokes the confusion of life and work separated only by a comma, and where the technological attractiveness works like a mirror for geeks. These messages forcefully create phantom links between professional activity and personal interests; creating, for example, promotional phrases like: "I participate in a project that develops software for satellites and I am an active member of an association that help disabled people fly on ultralight aircraft (ULM)..."[9]

Another aspect of the prescription of subjectivity lies in the attempt to artificially and forcibly re-establish a work ethic. In definitive rupture between management and the "troops" (this is the most commonly used term) cynicism, aggressiveness and competitiveness frequently emerge with episodes of bullying, harassment and discrimination. Lobbying and the practices of influence are management's daily bread, without even mentioning the cases of corruption that come to light in major international projects. In order to counterbalance appearances (and certainly not the reality), even IT services companies promulgate pompous "Ethics Charters" that describe the "moral values," rules for professional conduct (like audacity, modesty, and respect for others) and even the company's "social responsibility." These "commandments" are persistently hammered into the intranet and resemble greenwashing, the widely-used marketing strategy where a company flaunts an opportunist interest in environmental responsibility.

8 Internal communication campaign of the Capgemini company (around 2011).

9 Idem.

Competitiveness, open space, profiling and other pathogens of cognitive labor

Internal competitiveness finds its origins in company mantras like "Up or Out," invented by one of the first multinational North American consultancy services and now adopted everywhere: if in a certain number of years one has not been promoted to Vice President (VP), one is fired. Since it is obvious that in a business of several hundred thousand people only a narrow minority can become VP, everyone else is at risk. Competitive drive is discharged on various levels: collectively, where business unit executives of the same group struggle to grab customers and contracts, and individually, where a narrow minority of young authoritarian and parasitical managers trained through a single mold to extract profit from the direct labor of project teams and accrue rent through capturing value on the network. The common features are arrogance, contempt and use of humiliation. In this underbrush, there is no shortage of xenophobic and discriminatory attitudes towards women, ethnic minorities, seniors and low-rank trade unionists.

A step further down the ladder, we find project managers. They are forced to accept large quantities of work that can be billed to the customer on one side, yet are overburdened with the administrative tasks of reporting first to financial and management departments as well as productivity control departments,[10] leaving them with no material time to contribute positively to team projects. They work with the constant fear of becoming the scapegoat for any financial losses and they are at the mercy of customers who are often despotic since they are subjugated to the same treatment, and often lose all daily contact with their teams. The traditional role of mediator played by the project leader is thus dissolved.

In France, one of the countries with the highest levels of stress, these pathogens integrated into living labor have been revealed publicly through the media due to numerous suicides, which are nothing more than the tip of an iceberg of working conditions that often generate suffering. To complete the picture, continuous restructuring

---

10 The measurement of productivity of IT projects is highly subjective and it is often produced through empirical calculations where the financial value of the project is measured in man/days multiplied by the unit price of the day. See in this regard the KPI (Key Project Indicators) entry in the glossary.

campaigns like the infamous *Time to move* program instituted by Orange (formerly France Telecom) that imposes systematic transfers from office to office, or even city to city, every three years, compelling large number of employees, both state (initially hired on by France Telecom) and private contracts, to resign. Today, suffering at the workplace accompanies privatizations: suicides multiply in postal services or state agencies, such as unemployment offices.

Next in our review of pathogenic agents, we find the "Open Space," i.e. large offices without walls, where employee connected to the network are crammed into a minimum living space without privacy. This is part of a strategy of physical constraint. In a best-selling pamphlet published in France, open spaces are presented as the panopticon of modern labor (Des Isnards & Zuber, 2011), since this spatial conditioning is added to the conditioning induced by using network tools.

This last aspect leads us to a brief digression on the paradox between open space and telework. For all those workers who, for different reasons, do not spend all their time in the open space, the physical constraints of the panopticon are replaced by those of an oppressive use of network devices that tend to permanently break down the space-time barriers that once divided work from the rest of life, as openly claimed in the previously mentioned marketing campaign. This climate of fear and tension is also periodically revived through individual assessment. The old patriarchal management strategy of evaluations, however, undergoes a decisive transformation with the introduction of profiling methods, calibrated for the individualization of every single worker. The robotic use of specific applications and the indecent obligation of self-assessment reduce the cognitive worker to a "skills package," to an interchangeable gear in the industrialization of online labor, a "disposable" tool.

This dehumanization of the main moment of contact within the hierarchy is exacerbated by the breakdown between management and software engineers. In companies that sell services and "man-days," this personal data is opportunely used in a specific context, but always with the ultimate goal of making employees obsessively responsible for generating income and keeping busy and, if necessary, to push them to voluntary resignations, which in some countries remain much less expensive and complex than layoffs. It is therefore necessary to pressure all those who are "in-between," i.e. the time between one mission or project and another, or alternatively intercontracts. The longer this period, the more the collaborator must feel the weight of an "undeserved" salary. Obviously for seniors, who are all those over forty-five according to HR, this prescription becomes the dogma of "the self-made man." Even in companies with tens or hundreds of thousands of workers, salaried employees are responsible for its profitability and therefore should not wait for the company to give them a job, but must work hard to find it, though often employees does not have adequate means to do so.

One last point that confirms the will of IT services companies to regain control of certain horizontal, self-communication instruments of the multitude is the great trend of corporate social networks, seemingly emphasizing the existence of a company commons. The corporate social network[11] progressively replaces Knowledge Management (KM), now in decline for two main reasons. The first is the growing reticence in management to divulge methods and procedures that can be exploited by the competition, even within the company itself. The second is a consequence of the rush to productivity for which consultants or project managers no longer have the time to insert the contents and information in the company KM, tasks that are usually neither incentivized nor recognized. In some companies, employee participation in internal social networks is starting to become an element of evaluation and thus can influence employee career tracks and bonuses (a variable part of the remuneration that can now reach up to even 20% or 30% of the total wage).

Trade union decline and precarious work on the web: the French example

In France, perhaps more than elsewhere, unions are weaker and weaker and low membership, which was around 10% in 2016, testifies to this fact. The main unions have heavy responsibilities in having permitted and endorsed the degradation of the conditions of wage labor in general, and cognitive labor in particular. One of the first nails in the coffin in the long agony of union representation was driven in by the CGT and the CFDT, the two main unions who signed a 2008 suicide agreement with the neoliberal right-wing. In exchange for the consolidation of a presumed hegemony, the CGT and the

11 Social networking platform restricted to employees of a (large) company.

CFDT accepted a law that substantially reduced union presence in large companies, eliminating minority rights through drastic electoral rules. Then come the labor laws of 2016 and those of 2017, promulgated by ordinance, which, among other things, will once again affect union representation within companies and facilitate dismissals.

Despite some significant exceptions, becoming a union representative and remaining as one for a long time is often a pseudo-profession of patronage for salaried workers with a bent towards bureaucratic function and/or those in need of protection. Today, the role of the trade unionist is often reduced to managing, in an opaque and sometimes corrupt way, large budgets extracted from member dues and donated to "tourist-recreational" activities managed by *comités d'entreprise* [company committees], veritable companies within the company. The numerous embezzlement scandals in public groups are proof of this, like in the case of EDF (Electricité de France), the RATP (public transport in Paris), Air France, etc. A minimum of vitality at least remains in the battle against sexist discrimination in the workplace.

The role of the trade union as a containment mechanism is now quite reduced, especially in network enterprises and IT services; political initiative is, in fact, always in the hands of management, supported, in the rare instances of disagreement, by Labor Ministries aligned with employers' demands, regardless of the political color of the government. Often unions are content to fight for minor or marginal advantages in order to avoid disappearing permanently, especially when previously gained benefits are at stake, such as a reduction in the number union delegates, or union guarantees and prerogatives.

At the same time, in France there is the spread of precarization that affects young undergraduates or graduates with fixed-term contracts and the extension of the periods of apprenticeships and internships. These paths are filled with further obstacles, in the case of candidates of non-EU origins. The education system is constantly adjusting with this precarity: long-term work-study agreements allow the student to obtain quickly worker status, but with a low salary and all the work and study load. Engineering courses and training courses for a university degree or equivalent are increasingly organized according to the "needs" of large companies in terms of profitability and the very functioning of the *écoles d'ingénierie* [engineering schools] has been completely privatized. Despite this, unions continue to favor stable employees and are fighting to make the frontiers aseptic and impermeable between the company and the outside world: on the pretext that precarious workers and apprentices do not vote for internal union elections, they ignore the problem. It is a vicious circle in which the traditional union no longer appears able to fit into a context of producers who, both for their young age and for their social upbringing, have no culture, practice or background in "traditional" labor struggles.

Despite a series of trade union defeats and withdrawals and in spite of Macron's new political will to submit to neoliberalism, substantial social cushions still remain in France when compared to those of other European countries like Italy. Even so, we cannot forget that in France the pressure for hourly productivity, one of the highest in the world, is terrible and indelibly marks generations of digital cognitive workers. The attack against social welfare that was gradually implemented during previous decades is now abruptly accelerating. Unions are powerless to assist.

Burnout and escape

Schizophrenia animates the IT consulting sector: if, on the one hand, fences and network control tools are built, on the other, these companies must keep all hatches open in order to use a maximum of open source and free software. One example that rises above all is the enormous diffusion of Android, the Google operating system for smartphones and tablets – a derivative, like many others, of the flagship of free software, i.e. GNU Linux.

This schizophrenia is reflected in the work environment: to maintain profit levels, SSII (French IT consulting companies), must protect their "fortress" by occupying every inch of employee life and maintain a climate of stress and competitiveness in spaces created precisely for this purpose. Thanks to union support, SSIIs impose the forms of subjective prescription but, at the same time, must attract the largest possible number of candidates with an adequate technological profile. In this sector, other important factors play a role in the subjugation of the workers: job insecurity, which takes the form of temporary labor and project contracts; and, finally, the wage structure in which a variable component that depends on productivity and, above all, results becomes preponderant.

It is also important to note that SSII rationality becomes the model that financialized capitalism imposes on the public sector in order to implement the transformation of public welfare services into private outlets, thus creating huge, new markets to conquer. The system is trying in this way to feed itself: the destruction of welfare creates a social climate of fear that is conducive to accepting the pathogenic requirements of large private companies, and these conditions are applied in the privatization of public services. Thus the loop is closed.

Do the cracks from which new common creations emerge in the space-time of a metropolis that has become a cognitive factory of precarious workers or in the smooth surfaces of alienating hi-tech neighborhoods? How can we identify the contradictions that mask breaking points? What forms of struggle can be put into practice? It is hard to imagine the creation of ex-nihilo episodes of collective organization in SSIIs similar to those that inspired the autonomously run hospitals in Greece or Spain. Likewise, the strike as a form of efficient struggle is unlikely when the space-time of labor is completely unstructured. On the other hand, schizophrenia is not synonymous with stability and generates crises: the signals of stormy times ahead are beginning to appear.

The first is escape. Starting with previously mentioned authoritarian and parasitical managers, the majority of employees are driven in the first fifteen years of work to frequently changing positions, encouraged by management that offers stagnating wages. Later, once workers have reached forty, they resist in the vain hope of not being fired too soon! The more or less compulsive passage between one employer to another is also indispensable in obtaining a minimum of income progression but doesn't resolve the underlying nodes and risks. In the next company, the wage worker will rediscover the same rules and the same constraints. Obligated and resigned to a turn-over within a unified system of constrictive management, salaried network workers dream of startups as a liberation and sometimes obtain it.

Beyond the capitalist myth *à la* Steve Jobs, i.e. rapid and disproportionate enrichment, salaried employees seek a way out of a system in which they do not believe and that does not provide any hope or future. Participating in free software, many understand and appreciate the values of self-organization and cooperation. The attractiveness of peer production, free from pathogenic and compulsive company behavior, is the driving force behind abandoning wage labor, even if the startup, once established, may find itself trapped in the same system. In an effort to escape, there is still the dynamic of searching for a different mode of production which correspond to the social relations that have appeared in the latest social and territorial movements. The formidable assertion of free software and open source extends to science, art, biology and many other forms of knowledge and creation.

A second clue comes from the fact that an increasing number of IT consultants are evermore present in the network outside of areas exclusively dedicated to work. In the beginning, it was only specialized blogs but, little by little, the authors have begun to take a stand on issues of net neutrality, against labor policies, laws and the extension of copyright. More recently, in France, this class of network laborer has risen up against liberticidal espionage laws that "legalize" unchecked networking spying, inspired by the US Patriot Act. Sometimes, these digital workers go so far as to denounce explicitly, perhaps in a humorous or caricatured form, their working conditions. Management does not underestimate the potential danger and are ready to fire and prosecute workers when they publish these materials on the web and on social networks.[12]

A third and final indication of trouble times is given by the continued existence of the Anonymous movement that, as evoked in numerous articles,[13] has become a port of entry into politics for geeks and hackers who want to take action. Anonymous offers new opportunities for debate and simpler, fluid and confidential action. Some articles published in politically non-aligned websites[14] speak of the birth of mass hacktivism, new forms of protest and the antagonism of networked multitudes even if, for the time being, objectives are often limited to the issue of net neutrality itself that seems more and more in danger under the attacks of Trumpism.

12 Many examples in several countries, including Italy, of layoffs following the publication on Facebook of information concerning the employer.

13 See, for example the article of anthropologist G. Coleman (Coleman, 2011) and G. Griziotti, D. Lovaglio, T. Terranova, *Netwar 2.0: Verso una convergenza della "calle" e della rete*, 2012, in http://uninomade.org/verso-una-convergenza-della-calle-e-della-rete).

14 See http://www.01net.com/editorial/553558/affaire-megaupload-une-chance-pour-la-cyberdemocratie.

HACKERSPACE AND FABLAB

New Battles Between Copyright and Common Production?

The first communities of "makers" or digital artisans arose at the beginning of this century and they combine cooperation practices derived from digital technologies with the Do It Yourself (DIY) tradition. Thus, a movement has been born which foretells a possible techno-social and geolocalized transition in the way of producing material goods. Contrary to the writing of hacker programs that can work together on the net in different places, each in front of their computer, maker activity usually implies a more "physical" collaboration in hackerspaces, makerspaces or fablabs where hackers and makers conceive, build and creatively use machines based on free hardware/software, like 3D printers and every type of instrument controlled numerically by computer (CNC), including old knitting machines...[1]

In 2005 the RepRap project (short for Replicating Rapid Prototyper) was born in Bath at the initiative of Adrian Bowyer. This project has the objective of developing a 3D printer capable of

1 In a Parisian Hackerspace a non-professional knitting machine was interfaced with a computer. In this way it becomes possible to reproduce on a sweater scanned drawings or photos.

producing for the most part its own components. Although not a Marxist himself, A. Bowyer wrote a presentation for the project entitled "Wealth without money":

> Karl Marx and Frederick Engels wrote in *The Communist Manifesto* that "By proletariat is meant the class of modern wage laborers who, having no means of production of their own, are reduced to selling their labor power in order to live." This diagnosis is essentially correct... [...] So the replicating rapid prototyping machine will allow the revolutionary ownership, by the proletariat, of the means of production. But it will do so without all that messy and dangerous revolution stuff, and even without all that messy and dangerous industrial stuff. Therefore I have decided to call this process Darwinian Marxism... (Bowyer, 2011).

Actually, in many hackerspaces, the most revolutionary aspect is the transmission of knowledge freed from the institutional canons and that implies independent production free from intellectual property and other forms of accumulation.

Free hardware is a changing universe in continuous expansion. The first brick was probably "Arduino," a simple co-controller that costs about ten or fifteen euros, originally designed in Ivrea[2] and, when used with new, low-cost components, allows the General Intellect to have an additional weapon in the battle for the control of technological mediation.

We are probably on the eve of a new turning point that touches material production, including biological production. These new devices connected to computers and networks allow, in theory, to dematerialize and decentralize a part of the manufacturing of existing articles and their components. In the end, we can imagine that a part of production, today more and more centralized and distant,

---

2 Arduino (inspired by "Re Arduino" the name of the bar of Ivrea where the initiators met of the project) was conceived by a group founded, among other things, by Olivetti. A made to relate to the historical, political and economic role of Olivetti as a leader of innovation and of a "progressive" capitalism during the '30s glorious "post-war period.

will become local again by distributing across territory and thus, according to some, prefigures a further industrial revolution. As quoted by A. Giuliani (Vercellone & others, 2015, 124) André Gorz, with a visionary spirit, spoke of a new way social production based

> on the possibility of interconnecting artisanal laboratories founded on the common across the whole world, to treat – as does the free software movement – software as a common good, replacing the market with what needs to be produced and how, locally manufacturing all that is necessary and even to realize large complex installations through the collaboration of several local laboratories. Transport, storage, marketing and factory assembly – which represent two thirds of current costs – would be eliminated. An economy beyond wage labor, money and basic commodities resulting from the common activity immediately conceived of as common, announces the possibility of a free economy (Gorz, Ecologica, 2008, 118-119).

However, even in this field, the democratization and accessibility of tools are not necessarily a guarantee of autonomy. The dematerialization of objects that can be manufactured into a file activates the creation and use of a new copyright[3] to obtain rent from local manufacturing, as already happens in the case of spare parts.

On the other hand, the hacker-maker movement has a tendency to apply the principles and practices of free software, therefore creating free or open licenses. Having a physical place to learn, experiment with collectively, generate and transmit knowledge gives a particular value to this cooperation. However, there is a great variability of rules and types of behavior in these spaces: just think that to use the denomination "fablab," one must adhere to the statute established by the Massachusetts Institute of Technology (MIT) and that to obtain the "FabLab MIT" label, one must undergo specific training and obtain certification. There are many organizational methods in this field

3 Regarding English legislation, see for example Article *A 3D Printer's Guide to Intellectual Property Rights*. http://www.taylorwessing.com/download/article_3d_printer_guide.html#.VVY1r_ntmko.

and there is a wide range of hypotheses and possible behaviors that we will later investigate.

During a debate organized by the *Maîtres Ignorants* [Ignorant Masters],[4] members of a Parisian hackerspace stated that a physical place favors the ability of the members who work there, making them active people who no longer fall into submission but become more demanding, and concluded that FabLabs thus become a way to boycott a system that exploits the uniformity of the flock. In these spaces, there is a veritable dynamic that is created where the global knowledge that flows from it is greater than the sum of its individual parts.

The development of a still-experimental manufacturing in FabLab and Makerspaces could have socio-economic results comparable to the cultivation and distribution of organic farming. Similar to agricultural Solidarity Purchase Groups, it is possible to organize low-cost self-manufacturing entities of simple manufactured goods such as spare parts or products requiring little assembly. This would avoid the need for expensive factory mechanisms like serial production, transport, distribution and logistics with the accompanying accounting and administrative procedures at each step. For the moment, these experiments are still at a stage that does not compete with the serial and robotic manufacturing of post-Fordism, nor with the survivors of Taylorism. This is partly due to the fact that 3D printers and other machines that do computer-aided design and manufacturing are more suited to small quantities and not very competitive with larger, industrial machines. Delocalized robotic serial productions cannot replace standard industry overnight, but many questions remain.

How long will it remain cheaper to buy an object manufactured in the thousands in a Chinese industrial complex and transported home rather than producing it in a "kilometer zero" hackerspace? What geopolitical or environmental events may arise under such circumstances? However, it seems to us that the main question remains related to the sociopolitical implications of a new manufacturing paradigm whose scope and limits are not yet clear. We have insisted on the physicality of this production capacity because this element is an essential part of the makers' movement. The problem concerns the effect that opening these spaces will have on society. Although they may be places of autonomous cooperation and knowledge construction, do they

4 The *Maîtres ignorants* will be further discussed regarding new forms of learning later.

risk remaining accessible only to nerds and technology fanatics? In respect to technological culture, will they play a role similar to that which Italian Social Centers played regarding culture in general and, in particular, politics? In what way can they become physical places for realizing the common?

Another central issue is thus the future of this new mode of production. In the book C. Anderson (former editor-in-chief of *Wired*) dedicated to makers (Anderson, 2013), the "new industrial revolution" consists in the birth of a new business that, although it started from a widespread, grassroots production against big multinationals, does not call into question the capitalist dogma of the centrality of financial control over life.

FIRST EVASION

Time Devouring Spaces

Among the innumerable paradigm changes in progress, one specifically concerns the role of the two main cognitive entities or rather illusions for those who believe in physics: Space and Time. The evolution of the relationship between Space and Time in the metropolitan territory is a central indicator of biohypermedia change. Speaking about Paris, the philosopher Michel Serres[1] observes how time devours space: working inside the congested metropolis. A comparable amount of time is spent in several European metropolitan areas that he designates with the acronym PaMiLoBru: Paris, Milan, London, Brussels. In a financialized system, only time matters and, even in Paris, space can be devastated. Testimony to this is the disharmony of many large modern projects: the infamous *Bibliothèque François Mitterrand* (BFM), with its four inanimate towers that are deteriorating before our very eyes, or imperialist architecture of the Ministry of Finance that disfigures the banks of the Seine in strident contrast with the beauty of the works built centuries ago and when space was still an important value. The fact of living in time is now confirmed by the terrible deformity of transportation channels which,

1 In the radio broadcast *Le sens de l'Info*, 14 October 2012 (France Info), the philosopher Michel Serres is interviewed by the journalist Michel Polacco about space and time.

when entering the city, wind between asphalt, concrete and horrible commercial zones; the perfect visual expression of junk consumerist squalor colored only by the vulgar and chaotic writing of billboards. Territorial aesthetics are only a function of real estate rent. Even in sublime spaces, a stereotypical economic activity tends to render that space aseptic, uniform and impersonal, sucking the soul away and "Disneyfying" it. Public money is used to spit shine certain "tourist" sites or make them so. The result is clear in the historic centers of European museum-cities whether it is Venice, Prague or Bruges. Compulsory paths are created to move from one attraction to another like in Venice, where one is practically forced to go from Rialto to Piazza San Marco, with the façade of the Palazzo Ducale completely covered by an immense billboard of expensive watches. All these centers are standardized and rendered sterile by the same signifiers of "luxury." At this point, perhaps Las Vegas is more interesting because, in addition to casinos, you can see Paris, Monte Carlo or Bellagio and much more all in one place.

The obsessive search for capitalist rent requires us to live time differently, to live "of" time rather than "in" time. In the paragraphs concerning real time computing we have already discussed the emergence of a pervasive and regulated management of time associated with computer-machines. In the following section, "Living," we will see how the same phenomenon of time regulated by algorithms also extends to the metropolis as well, having replaced the factory and become a unified place of life and work.

Stephen Hawking inaugurates the Chronophage *clock at the University of Cambridge. Opened in 2008 with a presentation by Stephen Hawking,* Cronophage *is a monumental clock, driven by an electric motor with a 250-year autonomy, and mounted on a wall at Corpus Christi College, Cambridge (GB). Photo by J. Nathan Matias (CC BY-NC 2.0).*

Part Two:
LIVING

Bodies, perceptions and feelings in the space-time of biohypermedia companies

> Contemporary embodied subjectivities operate under a double imperative: they must be responsible for their surplus value, on the one hand as biogenetic containers and, on the other, as visible goods that circulate in the global media circuit and the global financial flow. Bodies today suffer from dual mediation: biogenetic and computer.
> (Braidotti, 2014, 128)

A second term of a dualism now opposing it to work, life is invested with such pervasive technological mediation that every minimal space seems to be progressively penetrated. Although such technological mediation has always existed and has played a disruptive role in some chapters of history, one gets the impression that never before has it affected time, space, bodies and subjectivity in such an increasingly inextricable and indistinguishable way from life itself. Although the purpose of this section does not purport to be exhaustive, nor to enter into new theorizations, we have still chosen to address technological mediation in relation to a limited historical, social and political context, however sufficient to establish some essential points of reference. Once a general outline has been established (since we cannot exhaustively address all the varieties of technological mediation), we have chosen to draw attention to some particularly significant and widespread issues, for example that of the interoperability of bodies and networks.

Historical context and global risks

The refined neuroscientific techniques applied to the economy are today more effective than Adam Smith's old concept of the "invisible hand." They draw inspiration from theories of a "selfish by nature" gene (Dawkins, 1979) and, above all, in the "historical demonstration" that any ideology not centered around economic rationality is doomed to failure. Before going into the aspects related to the technological mediation of subjectivities, we cannot avoid briefly mentioning the historical context in which this change takes place. The sudden implosion of the Soviet regime in 1989, whose importance was proportional to its area of influence and the impulse generated by the revolutionary influence of 1917, certainly gave a decisive push to the consolidation of neoliberalism.

There are no more witnesses of the long gone "ten days that shook the world" (Reed, 1961). But those who participated in the vast global movement at the end of the '60s that, on a geopolitical level, helped defeat imperialism in Vietnam, can remotely imagine the incredible strength that those events must have exerted over the subjectivities of that time. The fall of a now sclerotized Soviet Union was used to reach the conclusion that any attempt for any social organization not based on exploitation is only a dangerous utopia that favors the emergence of an autocratic political class. Without entering the debate, we could quickly argue that no revolution "wins forever" and that, by 1989, the Bolshevik revolution had already died half a century before. As correctly analyzed by Hardt and Negri, Stalin had already betrayed this revolution with the idea of nationalist socialism, which for "tragic irony ends up looking like... national-socialism" in Europe (Hardt & Negri, 2002, 115). The so-called Stalinist theories of socialism in one single country "reterritorialized" the communist revolution by putting national sovereignty (and, in this specific case, Russian sovereignty) at the center of everything.

From the point of view of subjective behaviors Vassili Grossmann, in *Life and Destiny* (Grossman, 1984), describes better than most the

Stalinist ideological perversion which, in the name of the good, induces terrible crimes and catastrophes. The dominant political class in the West, however, didn't waste the opportunity of using the implosion of a simulacrum to demonstrate that only capitalist cooperation is an effective and lasting system. With the complicity of mainstream media, the end of the USSR remains impressed in the collective unconscious as a historical failure and an obstacle to building a post-capitalist system on a basis that certainly can no longer be what it was for the communists of over a century ago.

The narrative of power drives even unorthodox thinkers to describe the period from the end of the Second World War to today as a golden era in the history of humanity! It is argued that the progress due to technology has enormously extended life expectancies, drastically reducing child mortality and that, despite "peripheral" wars, this era is, all in all, statistics in hand, much less bloody than the previous one. We are reminded that the real massacre is caused by road accidents where the number of deaths is much higher than any contemporary war or military operation, and we are reminded that about a century ago more people died every day in the trenches or in the great battles, than the number of victims of 9/11 in New York. This is an ambiguous thesis that supports the one circulating in the European social democratic circles: the current system with all its faults is ultimately the most acceptable!

This vision must be inserted in its historical, political and social context. In the "Glorious Thirty," struggles on many fronts, from labor to anticolonial movements, not to mention all the social liberation movements (like feminism, LGBT rights or antiracist struggles) were fundamental for global, social and political achievements. It is above all thanks to these movements that gains in the welfare state were obtained and the social use of technological advances are what have allowed, in many countries, improvements in length and quality in life. Later, after the neoliberal coup preceded by the fall of the Bretton Woods agreements, warning signs multiply. The rush of a financial system based on debt, which places public institutions at its service and becomes governance on a global level, is the basis for the continued widening of the income gap, of the increase of geopolitical tensions for the control of fossil energies which generate profound destabilization and uncertainty surrounding the irreversibility of social and ecological degradation.

Here we are discussing not hypothetical theories but tragic, tangible realities: in Greece, for example, the austerity cure imposed by finance through Troika technocrats gave rise to an "increase in child mortality of 43% since the beginning of the crisis," according to research published in February 2014 in the British medical journal *The Lancet*. Based on data from the Ministry of Health, cited in a *Washington Post* investigation that very same month, the number of abandoned children has increased by 336% in five years."[1] These are all elements that converge to designate the risk of an irreversible global crisis, incomparable with 20th century world wars or other older, but still limited, barbarities.

Technologies and the production of subjectivity

In response to this reactionary context, two major global waves of social movements[2] have already crossed the multitudes in this new century and have broken, leaving signs of a consciousness and knowledge which constitute a first level of change. Although they have provided general indications and instigated alternative practices, they have not, however, sedimented any political organizational structures that are much different than those of the past. The only exception could be Spain, where social movements have managed to take root, and the *Podemos* party has achieved notable success in local and national elections. For any electoral hypothesis of this type, however, the humiliation imposed on Greece in the summer of 2015 weighs heavy, with sanctions from the hegemony of world finance and German *ordoliberalism* definitively enforced in Europe.

At the same time, the European Union is the weak link in the global chain of power and this weakness in its institutions causes them to collapse in the face of national egoisms and the great exodus of migrants and refugees that many European governments helped create with nefarious and warmongering policies. The countries of Southern

1 F. Soddu and aa, "*Greece. Wounded youth. The crisis as a war, the country at a crossroads*," Caritas Italiana.

2 The two waves are 1) The altermondialist movement around 2000 and 2) about ten years later that which includes the Occupy and Democracia Real Ya movements (Indignados) in the West and the whole of the riots and revolutions of the Arab spring.

Europe are designated as the sacrificial victims for the strategy to standardize precarity conducted by financial governance. Faced with the breadth of this attack, even in progressive circles, a somewhat fatalistic resignation emerges. There are those who consider inevitable – be it for cynicism, impotence or perhaps both – the historical nemesis of a Europe that must pay for its imperialist past through the downgrading and the precarity of a third of its population, forced to live with four or five hundred euros a month. It, however, must also be stressed that this would be a dream income in the favelas of Rio, in the Soweto slums or in the textile factories of Bangladesh!

Beyond the disturbing reactionary and xenophobic populist epiphenomena that thrive by exploiting the desperation of the most vulnerable and an expertly maintained climate of insecurity, how can the apparent inertia of the multitudes of the Technological General Intellect be explained? It does not seem that this passivity is due to a widespread and profound adherence to capitalism, in spite of the means this latter has at its disposition, including the media. The hammering traditional media – radio and television – during the Greek crisis amply highlighted the existence of an uncritical mass that supports financialized governance. In this situation, the decades-old myth of an alleged "objectivity" in "independent" news sources claiming the separation between facts and opinions, definitively collapsed.[3] On another level, for certain countries, government agencies are tasked with making the new generations believe that the context of total precarity and ruthless competitiveness is "fun," using the mirage of Californian-style startups.

In this section, we will try to analyze how these mirages function and what role the main techno-algorithmic and application devices implemented by digital corporations and entertainment play. It cannot be denied, for example, that the functions proposed by social networks initially corresponded to a need, and this would at least partially explain their success. On the other hand, it is no secret that services like Facebook also function as attention traps and encourages egocentricity. These aspects are part of a general framework where the capacity of cognitive capitalism to use technological mediations

---

3 This is the case, for example, of *Le Monde* considered in the past as everyday French reference for its "objective" line and now completely aligned on positions of neoliberal orthodoxy.

over time for the genesis of new subjectivities is evident. The implicit goal is the structural integration of the free market's economic form directly into subjectivity as the only and all-encompassing model for all social, economic and political relationships, including individual behavior.

Let's start from the principle that this hypothesis of the exercise of influence and manipulation, while partially true, is incomplete and needs to be further explored. Perhaps it is not simply a question of "mediation" in the sense attributed to this term until today: technology has, in fact, broken what are considered intangible barriers and has now become part of living matter. In this case, too, we enter a field in which capitalism could have a "genetic" advantage. As Foucault claims, in fact, liberalism was born in the 18th century from the "*laisser faire*" principle, as opposed to the disciplinary system that existed previously (Foucault, 2005a, 198-199). According to the French philosopher "the motor of action is desire [...] and the naturalness of desire that characterizes the whole population can be shaped by techniques of power and government" (Foucault, 2005b, 74-75).[4]

It can be deduced that the ability to manipulate desires in the political sense, giving life to consumerism among others, is part of the capitalist genetic makeup, just like an obsession with property or accumulation. This "innate" characteristic, however, does not necessarily imply that a technological "heart of darkness" of contemporary necropolitics, symbolized by the devastating use of war drones or the potential destruction of nuclear power plants, is inevitable.

On one hand, there is the apparent neoliberal advance in the use of technique, and capitalism's equally genetic mimetic capacity to adapt to paradigm changes; on the other, a widespread drive to break this spell, to resist manipulation and to escape from the "Truman Show" effect (Weir, 1998) is developed in "crisis" countries.[5] Behind what may seem like incomprehensible inertia, there is a spread of initiatives, experiments and alternative attempts that are possible today thanks to autonomy and knowledge. To counteract these actions (which are, for

4 Foucault, 2005b, 74-75.

5 The film tells the story of a man, Truman Burbank, the protagonist of a reality show without his knowledge. From birth his world is a giant television set and all those around him are actors. He alone ignores reality. The film presents his first doubts and his search to discover the truth.

the moment, minority or perhaps simply "underground"), the system rapidly intervenes upstream, introducing elements of divisive individualism, creating enclosures around which control and the extraction of value are exercised and, in the last instance, when the waves finally do rise, react with harsh and "surgical" repression.

Our hypothesis is that automatic control can be unraveled through the development of autonomy, and that the capitalist technique of the anticipation and manipulation of desires can become a relative value in the framework of an expanding grassroots trend that can no longer be contained or fenced. Perhaps in this very moment a cyber-rat is digging deep through hypermedia even if we do not know exactly which way they are is going. The match is uncertain; it is running on the edge of an asymmetrical razor.

BODIES "FORCED TO BLEED"[1]

The '70s: Disobedient Italy

In the hypothesis of global risk, subjectivities are at the center of attention. In an even more abstract and indirect way than the US drone operators that kill tens of thousands of miles away using video game joysticks, the aseptic financial decisions are reflected in time and space, and in bodies. It is a common practice of governance to make key decisions in particular moments when social reactions are weaker. It was in a torrid August 1971 that Richard Nixon announced the end of the US Dollar's convertibility to gold, a policy established in the Bretton Woods agreements in 1944. Since then the United States has been able to freely print currency and to develop and strengthen their political and economic command over the world.

From that August, the face of the world economy changed radically: Nixon's decision was a real revolution that strengthened the role of the United States in the global economy, laying the conditions, or premises, for their international economic supremacy. As long as the Dollar remained anchored to gold and other currencies, the US had to respect that parity and could not arbitrarily distribute their currency around the world, generating inflation. From 1971 on, however,

1 Title of an essay on the punk movement in Italy in the '80s (Philopat, 2006).

free from any monetary constraint, the United States has been able to finance rations abroad, simply by printing money. They have thus expanded their economic and political influence to other continents.

This is probably the first conservative counter-attack against the social movements of the "Glorious Thirty." An attack that finds its political implementation a few years later with the rise to power of ultraconservative groups in the two main Anglo-Saxon countries, guided by symbolic characters: Ronald Reagan, a former actor and McCarthy-era trade unionist and Margaret Thatcher, the London suburb shop owner that would make grit her distinctive political sign. From that moment on, value curves of labor and rent began to reverse: the former declined and the latter went on a rise that has lasted until today. In years from the end of Bretton Woods to the Reagan election, a general reflux of the global movements and struggles of the past decade took place, while neoliberal policies were affirmed. Only Italy resisted and social movements were in full swing until at least 1977.

During these years, the capacity to understand instinctively the functioning mechanisms of capitalist society, a capacity induced by labor struggles and the internal productive cycle, emerged within the subaltern and productive classes. This therefore led to the creation of a new consciousness and subjectivity. At the same time, restructuring must move forward and cannot bear the risk of an irreversible rupture with capital caused by a mass intellectual capacity linked to the avant-gardes and no longer under the control of unions or the Italian Communist Party (PCI). Hence the enormous repression in Italy which, first and foremost affected the avant-garde, but was also directed against the masses, showing how dangerous it is to "break rank." The direct role that the PCI and unions played in the scientific and systematic destruction of those subjectivities is now in the annals of history.

Our purpose here is certainly not to analyze the reasons for this Italian peculiarity, but rather use it as a reference for analyzing the political use of abnormal instruments by biopower. In a fateful autumn in 1980, Reagan's election intertwined with Italy through the "March of Forty Thousand White-collar Workers" at Fiat in Turin, a moment that represents the strategic defeat of the unions in the face of Confindustria.[2] This is the prize that the union obtained from the capital for having made the decisive step in criminalizing antagonist social movements.

From that moment on, the capitalist offensive spread, moving the epicenter of production and labor struggles from the factory to a fragmented metropolitan territory. As a result, the working component of the youth proletariat, which had become a central political subject by the end of the sixties, was expelled from the factory and found itself lost in the metropolis. The swan's song of that movement occurred in two different places: first in Milan at the last Lambro Park Festival[3] in 1976 and in Bologna at the 1977 Autonomy Conference. This despite the fact that those who participated in the two events had the feeling that perhaps, for the first time, the political themes of Organized Worker Autonomy had become "mainstream" for a whole generation...

Going over the documents of the time, for example the video material presented by Alberto Grifi (Grifi 1976),[4] one perceives the great creativity of a movement that in some ways already prefigures post-capitalist ideas and behavior. At the same time, the movement had probably peaked. Although globalization was not

---

2 The march of the white-collar workers of FIAT takes place in Turin, October 14, 1980 to protest against the pickets and the strike proclaimed by the unions in response to the "political" layoffs and the setting up of "Cassa Integrazione" of more than twenty thousand workers by the management of the company. The march that actually supports this is considered a breaking point in the history of trade union struggles in Italy. For the first time in the history of workers' protests the so-called "silent majority" of the most important industry in Italy consisting essentially of cadres, employees, chiefs, caplets and other skilled workers raised their voice and gave rise to a demonstration that would forever change worker, union and company relations. The demonstration would represent a hard failure for the unions.

3 In 1976, from 26 to 30 June, the last, troubled, was held at the Lambro Park edition of the Festival of the youth proletariat, in which more than four hundred-thousand people.

4 See the amazing video shot at Lambro Park by the militant filmmaker Alberto Grifi thanks to a collective of about twenty people with new "popular" VHS cameras. The Milanese festival took place in an atmosphere of revolt against the organizers themselves but also with the creativity and reflections of a Post-revolution.

at today's level, it is the "flight forward" of a single country that could not continue indefinitely. Within the movement itself centrifugal and contradictory forces, from armed extremism to the mystical flight to the east were becoming ever more preponderant. Immediately after 1977, in an already deteriorating climate, the coup de grace came in the tragic form of the abduction and murder of Aldo Moro. An entire generation that participated in the long run of the Italian movement – born in the hot autumn of 1969 – paid the price for its resistance and its final virulence with a dramatic national lobotomy that definitively excluded it from Italian political life.

After the apogee of 1977, a period in which the first signs of reflux, the *longa manus* of repression weakened student resistance and the social workforce in the meshes of urban fabric through two main local agents: police and mafia. To tighten this grip and destroy any ambition of revolt the body becomes an important tool in these biopolitics. In the eighties, in Italy as well as in many other countries, drugs (above all heroin) were used as a biopolitical device that acts on bodies to stun them, paralyze them, make them dependent and neutralize them.[5] This was the explicit goal of *Operation Blue Moon*:

> an undercover operation carried out by the secret services of the countries of the Western Bloc in the early 1970s, in the context of the Cold War, aimed at spreading the use of heavy drugs, particularly heroin, among young activists of youth movements of protest, in order to make them addicts and divert them from the political struggle.[6]

The mafia acts in connivance with the institutional organs of

---

5 See the *1977* interview*: Eroina e Rivoluzione – Conversando con Vincenzo Miliucci.*

6 In http://it.wikipedia.org/wiki/Operazione_Blue_Moon. See also the historical reconstruction of the activity of the "Nucleo Antidroga" of the Carabinieri of General Miceli, later accused of an attempted fascist coup, in the book by G. Blumir (Blumir, 1983) and in the RAI document by A. Ricucci, G Foscarini, *Operation Bluemoon – Heroine of State*, The story is us – Archives Rai https://www.youtube.com/watch?v=kywmDZVjTnw.

repression and through the networks of drug dealers in those years linked to the neo-fascist areas."[7]

The method is to make light drugs disappear or make them very expensive, while the mainstream media launches a repressive campaign of disinformation on the use of the same: for example, putting heroin and cannabis on the same level, i.e. equating "smack" and "weed" as "dope." At the same time, large quantities of cheap heroin are put on the market, using a complementary and longer-lasting market cycle to encourage involvement towards increasingly heavy drugs. Between 1972 and 1980, these cycles begin with low cost amphetamines and, little by little, intravenous assumption methods ("shooting up") were adopted. Later, access to the morphine was granted, only to be followed by the definitive step towards heroin.

This shows how the subjectivity of a generation was dangerous and how, beyond the main tools of repression like, for example, "exceptional" laws (that are still on the books), it could be weakened with heroin, just as it had been used in the black ghettos of Los Angeles in the fifties and sixties. This biopolitical attack is an important junction that leads to the progressive confluence of financial capitalism with the mafia and organized crime. Already at that time, beyond bombastic declarations, neither the world of finance nor the political world had any moral scruples in using mafia money whenever possible or necessary, to the point that, in the opacity of markets and transactions, the distinction between legal and mafia funds is blurred and confused. In this specific case, we are not talking about episodic or anomalous collusion, but a precise political will to execute a specific action.

Biopower, in addition to the classic institutional Foucauldian settings, schools, barracks and prisons, is exercised on the bodies captured in the metropolitan territory in a subtler way. With heroin, it is no longer just about modeling bodies, but about annihilating them if they are no longer useful or functional to the system. It seems that even the then-president of the Italian Republic, F. Cossiga, spoke of using state heroin as a necessity for Italy, the last rebel and recalcitrant country to capitalist normalization, to recover.[8]

---

7 The living forces fighting this biopot attack on the ground were the first social centers and organizations in the Autonomy area. It was in this and for these reasons that the fascists murdered in Milan in 1978 two eighteen-year-olds: Fausto Tinelli and Lorenzo "Iaio" Iannucci.

8 On state heroin see also the film *Slow.* (Church, 2004).

One wonders if today, a few decades later, we do not find ourselves in a similar situation, even if there are no "rebel" movements to tame. The central pivot is no longer heroin (whose use is again growing), but of the most widespread and "tolerated" of drugs. Today's derivatives of *cannabis sativa* have a concentration of the active ingredient THC that is much stronger than in the past. Although obviously liberalization is the only acceptable option, one wonders if it is not a coincidence that only now the bastions of repression are beginning to crumble. Not only is it a great lucrative trend for multinational tobacco companies who are currently preparing for widespread commercialization, and for states that are set to make a lot of money through heavy taxes, but it can also be seen as an adjustment in the parameters of biogovernance. The first consumers are young people who, in many European countries, live in a state of unemployment and precarity. In this situation, a "liberated" and strongly taxed drug with ability to appease is likely to be an adjuvant to passive acceptance of an alienating situation, just like heroin was an instrument used to accelerate the reflux of social movements in a previous era.

Psychotropic medicines are also the "clean" drug of choice for capital because, in addition to carrying out the same function as cannabis, they already make huge profits for "big pharma" and its sponsors, not to mention video games...

The system of allergies and autoimmunity

An apparent and tangible confirmation of the above comes from the role of the body as a central element of expression in recent generations. From the punk generation on, tattoos, piercings, even scarifications, seem to be signs of a biopolitical impact on the body. The attack of biopower leaves ever more extensive, evident and perceptible traces on our skin, our flesh, as well as in our psyche. Punk artists have denounced this attack in their own way since at least the '80s, like the book which gives the title to this chapter states; aren't changing our appearance and deforming our flesh not reactions to the suffering imposed on the body by a triumphant power?

On the other hand, the bodies become more and more sensitive, susceptible and alert to the environment and lifestyles. It is not our goal here to explore the theme of the capitalist degradation of the environment, but we will highlight how the allergy explosion and autoimmune diseases[9] have implications that go far beyond the problems induced by pollution and urbanization. Allergies are multifactorial and depend not only on genetic causes but also environmental and cultural changes as well.[10] They represent a way the body defends itself from daily attacks, not only on a physical plane. For example, there are studies that document how the frequency of asthma and allergies has doubled in the last forty years (Annesi-Maesano & Baiz, 2012), and still others that highlight how a stressful situation experienced by the mother increases the risk of asthma in the child. There are more and more people living in cities where we are increasingly intolerant to the polluted air we breathe, both outside and inside and where we eat fa(s)t-food. The stress generated in a context of tension, of competitiveness, absolute uncertainty and exclusion aggravates allergies. And like allergies, autoimmune diseases are on the rise.

These diseases may be systemic or specific to an organ, but they have in common the principle of a malfunction of the immune system that attacks their body structures. According to researchers many objective factors may contribute: degraded environmental conditions or, as some argue, increasingly disproportionate utilization of medicines, encouraged by the pharmaceutical industry, in particular anti-inflammatory and antibiotic molecules, but also subjective factors are certainly present. The fitting definition given by the French national medical research institute Inserm is: "autoimmune disease is a rupture in our tolerance of ourselves."[11]

Isn't the message of increasing intolerance of our bodies and organs when we are "socially" mistreated implicit in this dynamic? A further weakness of the species comes from diminishing male fertility, revealed by a French study published in the medical journal *Human Reproduction*.[12] As in the case with allergies and autoimmune diseases,

9 Among the best known: lupus, autoimmune thyroiditis, certain types of diabetes, multiple sclerosis, the Crohn's disease, rheumatoid arthritis, celiac disease, etc.

10 For example, allergies to dairy products are probably linked to the changed minds in ways breastfeeding of infants in Western countries.

11 http://www.inserm.fr/thematiques/immunologie-hematologie-pneumologie/dossiers-d-information/maladies-self-immunes.

12 The data covers a sample of over 26,600 men between 18 and 70 years, and was collected between 1989 and 2005 in 126 reproductive health centers in France. The number of million sperm per millimeter decreased by 32.3 percent, at an

researchers rule out the idea that genetic factors, whose mutation require centuries, might have caused the drop in sperm count of nearly one-third in less than twenty years. The experts cite continuous environmental and alimentary degradation and the indiscriminate use of pesticides in agriculture, all caused by the involution of our social organization. Faced with this situation, disciplines that deal with the relationship between the nervous system, the immune system and the endocrine system are being developed.[13]

One might assume that, when facing such diversified and invasive forms of control in the professional sphere, they go beyond the temporal and spatial limits of work, pouring over into life. Furthermore, a growing number of people might feel cheated by life itself, incapable of opposing any effective and collective resistance. Facing loneliness and surrounded by an individualism that is encouraged as a supreme virtue, bodies and organs react, rebelling or losing the capacity to reproduce. As we continue this discussion, we should bear in mind these factors.

Acting on the symptoms, leaving the causes intact

Bodies naturally rebel in their own way to a harmful environment, due to pollution or problems of a different nature. In the natural environment (which is certainly the most measurable and widely examined area – even if it is often denied or hidden according to the financial and political interests at play), there is also a neuropsychological environment that may be less visible, but no less invasive. This type of pollution is created through a situation of continual fear exerted by a government in a perennial "state of exception," by the extension of workplace stress into life, and by introducing more or less surreptitiously, elements of behavioral and psychological conditioning through particular uses of technology.

The response of global, national or local institutions to all of these diseases and illnesses that emerge in Europe in a diffused and striking way, consists in merely symptomatic or mediatized actions. The policy of precarity implemented in the name of competitiveness cannot but help create an environment of anxiety. This environment produces

---

annual rate of 1.9 percent.

13 E.g. the Psiconeuroendocrinoimmunology.

diseases that create an interesting "market" where containment activities and symptomatic care are developed. Though it is simplistic to focus only on drugs and increasing progressively their use, is it only coincidence that France, a country characterized by widespread workplace suffering, has the highest consumption of psychopharmacological drugs in the world? Other disciplines – from neuroscience to genetic engineering – are involved in acting on the bodies and minds. Finally, even if these aspects are more superficial and anecdotal, it is curious to note how a nervous society generates antibodies of all types and new sociological phenomenon: the proliferation of fitness centers, Hamman and thalassotherapy, the increasing popularity of oriental psychophysical exercises such as yoga, tai-chi etc., and the appearance in the western metropolis of a dense network of "relaxing" Asian massage boutiques. These can only testify to the increasing need to escape from stress and attempt to restore balance.

In search of the lost DNA

We hear a wake-up call given more and more often at the movie on the topic of the biological relationship between parents and children. Sometimes they are mothers who leave after decades looking for their daughters from whom they were separated after birth or, vice versa, children who go searching for their origins or, sometimes, parents who discover they have no genetic links with their child.

Through these events, screenwriters and filmmakers remind us of the importance of the biological link, as if, in a hidden way, they wanted to emphasize that it is in danger. The topics are varied, but always centered around a mystery or a search for a genetic relationship, even if the settings vary. The incident may concern the past, like the story of the Irish Catholic boarding schools in the '50s where children from single mothers are sold to rich American couples (Frears, 2013), or the drama of the children of the SS's eugenics program.[14] The scripts may relate to current events, like movies about hospital mix-ups in the maternity wards of hospitals which are sometimes only

14 (Mass & Kaufmann, 2012) Films dealing with the *Lebensborn* under the direction of gerarca SS H. Himmler to realize the eugenic theories of the Nazis and develop a pure, dominant Aryan race.

belatedly revealed by genetic testing.[15]

A "strand" of DNA binds these different stories. The acceleration of genetic engineering and related technology that radically changes human procreation generates a consequent loss of reference points that have always existed and that drive the collective unconscious. Artists, as we know, are particularly sensitive to the hidden signs of warning and distress, which they often make visible in their works, that are born from the fear that the steamroller of commercialization will also devastate the biological aspects of life.

15 In the Japanese film *Like Son Like Father* (Kore-eda, 2013) and in an episode of Brazilian film *O que se move* (Gotardo, 2012).

THE TURNING POINT OF BIOHYPERMEDIA

Cybernetic and robotic machines, information technology, cognitive sciences, biotechnology and nanotechnology hybridize like fractals in order to come into intimate contact with biological life, to the point of becoming indistinguishable. Figures like the android, the neuromancer and the cyborg are born and the interface between machine and bios, after being worn so thin, eventually dissolves into a continuum. In this new environment, only an interpretive prism that can integrate this paradigm shift will allow us to understand the dynamics of behavior, attitudes and social structures.

It will be, however, an interpretation obstructed by the continuous tensions of a system that renders everything calculable. Are algorithms able to mold affects without transforming them into something else that exist? Or are we already at in a posthuman state where algorithms, desires and affects are inextricably bound? And in that case, what about ethics? This situation obliges us to undertake a symmetrical process to the one digital corporations exercise in networks, capturing knowledge, anticipating and manipulating intentions.

In this regard, we will attempt to examine the critical points in the maze of algorithmic and technological devices of biopolitical control. In order to reduce the complexity of the case, a few different subdivisions may be useful, if not necessary. Such subcategories include:

- Analyzing where technologies and algorithms resonate with emotions, feelings and affects in order to identify what impacts are generated and where these neurodigital processes aim to transform desires into impulses;
- Deconstructing complex systems into subsystems that are more intelligible or controlled;
- Searching for processes that can subvert or destabilize these attempts at manipulation, or at least find possible vanishing lines.

Human and machinic awareness

We previously touched on the concepts of "interrupt" and "real time" in the early digital years. These are what, among other things, allow us to understand industrial machines piloted by what is called Computer Numeric Control (CNC). This means machine tools like presses, press breaks, punch presses, lathes, milling machines, welders, sheet metal cutters, etc. whose movements are directed by a computer (either integrated or external) that commands its movements and functions according to a program predefined through software. These machines are limited to executing the set program regardless of what happens around it and we should distinguish them from industrial robots, which have more autonomy. Operators are responsible for the preparation of individual pieces and must intervene in the case of unexpected events, thus requiring CNC machines to be operated always under human supervision.

A CNC machine can substitute both the manual and repetitive labor of the assembly line, constituted by the monotonous mechanical effort of the human body while the mind runs free, as well as that of the specialized laborer who makes complex pieces using lathes or milling machines and who has advantageously assimilated these skills, while working faster and with more precision. In a CNC factory, it is above all the operator's awareness that becomes an integral and central part of labor. Surplus value is therefore generated by the few trained humans left to monitor these automatic processes. The need to supervise diminishes even more with the arrival of industrial robots that can theoretically function autonomously and that are able to adapt to variations in the environment they operate in, while still completing the task they have been assigned. Thus, yet another part of human awareness required for supervision is transferred to the robot.

Interrupt, the principle that initially gave the computer the capacity to be aware of the world around it, is really a machine's ability to acquire knowledge to the point that, in a factory, the machine progressively incorporates the whole of a laborer's skills: physical force, technical ability and awareness. This evolution in awareness also relates to digital cognitive activities like programming. Towards the end of the '70s, there were few computing centers and the growing number of students learning to program had very limited resources at their disposal. In those years, for example, the Politecnico in Milan – one of the most important technological universities in Italy – only had one computing center with a single Univac 1108 scientific mainframe. Access was quite limited and the procedure for the few authorized students was long and complicated because even a simple program had to be run multiple times before completion, with each test requiring several days to obtain an output.

So, with the high cost of computers, a substantial part of programming pedagogy was dedicated to the proofing of orthographic, syntactic and logical code. Students were taught, in a nearly compulsive way, to check mentally instructions step by step in order to eliminate syntactic or logical errors on paper to reduce the number of times the program went through the computer, thus minimizing the amount of time spent in the computing center. Contrary to what happens today, human attention and concentration were abundant "goods" while calculation time was scarce and expensive. Only at the beginning of the '70s did the first minicomputers arrive from the United States, computers that were more affordable for the university and for mid-size companies. They functioned without needing the costly installation of a computing center. Twenty years later, when PCs became largely accessible and calculation time became much more rapid, the myth of the programmer's preventative concentration started to crumble.

In the case of both the factory and cognitive labor, we see a progressive delegation of awareness to the machine. At the same time, machinic presence multiplies in the workplace as well as at home, contributing to an increasing porosity. Diffused and omnipresent devices incorporate knowledge but also solicit continuous dialogue. The "background noise" of the machines that surround us has a Freudian "Es" function. Just like in the Freudian intraphysic instance, the machinic "Es" tends to become a continuous flux of solicitations and information that grab our attention and that are difficult to control.

This incessant stream is often made up of a mix of stimuli that concern both our personal lives and work, where each can hold a certain level of priority, pressure or tension. It is the management of this complexity and the fragmentation of our attention that characterize a space-time in which any separation between life and work tends to become ephemeral.

The cellphone, a biopolitical device

In dealing with new modes of production and with the emergence of the so-called New Economy in the first part of the book, we focused our attention on mobile telephone operators for the speed with which, beyond telecommunications engineering, these businesses accelerated the development of influenced technologies like telemarketing and advertising, in a sort of gold rush that lasted for at least a decade. In this section, the use of the mobile telephone is the object of our study, the "cell" that becomes a fetish and occupies an important space in everyone's lives. First, we'll discuss mobile telephones and "voice" cellular services as they were in the beginning, when they spread on a wide scale in the second half of the '90s; later we will address their transformation into smartphones and universal interaction devices.

Born from the French impulse that would bestow the name GSM (Groupe Spécial Mobile), later to become the global acronym Global System for Mobile Communications), the 2G standard would become the European norm and, in 1991, the first experimental communications took place. Today, the GSM network conceived above all for 2G mobile voice telephone communication is still the most widespread standard in the world: in 2015 more than three billion people in 200 countries were using GSM cellular telephones. Along with 2G, there are other services that make mobile telephones a global phenomenon, as we will later see.

Our hypothesis is that the mobile voice telephone service gave birth to a process of profound change in subjectivity and that, later accompanied by the internet, this shift became constitutive, together with other technologies, of irreversible anthropological transformations. In the Ptolemaic universe of ICT, everything revolves around the internet, but we should at least place the "star" of mobile telephone services beside it. Even in the initial period of GSM, the "cell" quickly became – and remains – the biopolitical technological device *par excellence*, if we consider it through the eyes of Foucault. What happens when such a pervasive technology bursts into the context of social relationships of production in capitalism? Here below we will set out the principle characteristics of this phenomenon in an attempt to highlight the effects and implications in the formation of subjectivity: universality, accessibility, attention, deterritorialization, sociality, control and productivity.

- **Universality –** The adoption and spread of mobile telephone service probably have unprecedented characteristics because it satisfies a new and ubiquitous need for communication which might explain the relative synchronicity of its global expansion. Currently, a near totality of the global population has access to a mobile device that, as a minimum, enables oral or written communication (text, multimedia messages, etc.). In less than 30 years, cellular phones became the biggest and fastest technosocial revolution in the history of humanity. Unlike print, automobiles or televisions, this expansion was nearly uniform and simultaneous, both in industrialized countries as well as emerging economies and less developed nations, allowing for a generational leap in technology with more access to communication for everyone. In many countries where the traditional telephone remains a luxury for urban elites, cellular communication through prepaid plans is accessible for practically everyone anywhere, at least in terms of reception, and top-ups are often available for a just few dollars in these countries.

- **Accessibility –** Mobile telephone service is relatively accessible and immediately skips over all other widespread technologies: in fact, it does not require important investments, nor does it demand a continuous source of electricity like television; it doesn't even need a stable home. Unlike the use of digital devices like the PC, mobile services are concentrated in a handful of simple functions and are immediately accessible to everyone, even to someone illiterate. In addition, they do not require any specific skill, do not entail any generational divide and do not involve any prior training.

- **Attention –** Communication with cellphones absorbs a large amount of attention from the user and demands a certain level of isolation in respect to the surrounding environment. Its use entails a step forward in our real time understanding of social communication:

> The cellphone is the ultimate compressor of social space because it allows real-time communication from any place with reception to any other place. The compression of time and space that the cellphone can handle is akin to a worm hole. The caller goes into a partial black hole of perception as a phone call is taken, when the caller connects to the call-ee, a wormhole forms in time/space, allowing communication to happen through the two individuals.[1]

Many of the studies on human attention while communicating on a cellular device were conducted to analyze the limits of our capacity to concentrate on something when our minds must manage two activities. The inattention of a person whom we come cross and who is speaking on a cell is perceptible, and this becomes dangerous when this person is simultaneously performing an activity that can harm others like, for example, driving a vehicle. For Cristian Marazzi, this attention hoarding can even put the New Economy in crisis:

> An economy innervated by communication technology needs a large amount of attention from consumers in order to digest the offer of goods and services. Since the New Economy is, in fact, an economy that consumes not only working time but also free time or non-working time (in the sense that all of life is put to work), it follows that the crisis in the New Economy is determined by the contradiction between economic time and free time. In other words, the crisis explodes

1 Case, 2008, 36.

> due to "excess of economy," the disproportion between cyberspace and cybertime (in Franco Berardi's terms).[2]

Starting with cellular phones, the interrupt-attention relationship exceeds the space of productive relations to arrogantly barge into private life:

> The Internet is an interruption system. It seizes our attention only to scramble it. [...] Every time we shift our attention, the brain has to reorient itself, further taxing our mental resources. Many studies have shown that switching between just two tasks can add substantially to our cognitive load, impeding our thinking and increasing the likelihood that we'll overlook or misinterpret important information. On the Internet, where we generally juggle several tasks, the switching costs pile ever higher.[3]

All aspects are concerned: relations, sensations, feelings and, in this context, we even find certain parameters of social control. The subjugation of attention through the use of technology makes it a rare type of good and therefore an economically interesting merchandise, as Tiziana Terranova appropriately maintains.[4]

- **Deterritorialization –** The cellular phone changes the social geography of public and private space and allows for direct and immediate communication from one place to any other. In this sense, cellular service paves the way for a deterritorialization that is later developed through technologies relating to mobile internet.

- **Sociality –** The use of mobile phones creates particular spaces between the sociality of intimacy, face to face encounters and

2 Marazzi, 2002, cover.

3 Carr, 2010.

4 Terranova, 2012.

the sociality created by conversing at a distance. Users are in a residual space in which they are no longer entirely here nor there, or one that sends them off searching for a private space (getting out of the car, leaving a meeting or going into the hall). On the other hand, the use of cellphones creates new forms of sociality. In fact, while communication expenses can initially limit use, when low-cost monthly plans are available, a new form of sociality is generated thanks to a stable and open channel that becomes a surrogate for being physically present.

- **Control –** A counterweight to the relational potential enabled by cellular communication also rapidly becomes an instrument of personal biopolitical control.

- **Productivity –** At the beginning of its expansion, the cellphone was quickly adopted not only in all the nomadic professions but soon afterwards also by the majority of workers. Although personal use was privileged, in the climate of stress that more often than not characterizes cognitive labor, it becomes a tool for productivity and control in the hands of managers, even if it forces them (and not only) to broaden their availability and their surveillance function over space and time. Here we are facing the first technology to directly and continuously break the separation between work and free time. Only later will the use of cellphones be imposed on precarious workers as a tool for exploitation and self-discipline.

These brief considerations on a handful of the aspects of cellular voice services provide a few indications on the influence that they exercise over subjectivity. We are dealing with an individual vehicle that, at the same time, inaugurates the P2P era. Introduced into a system that exalts competition as an absolute value, the mantra of global governance, the cellphone is the means of communication of individualism. However, it also lets us, in as much as the first P2P tool, to move more easily beyond barriers and hierarchies and is therefore an instrument of horizontal communication too.

Criticism of "becoming-machine"

The concept of biohypermedia, understood as the continuous interaction of bodies and organic nervous systems with the world through the complex of mobile devices, applications and reticular infrastructures, was introduced in a special edition of Alfabeta2 magazine, entitled AlfaBiohypermedia:

> Biohypermedia is the current dimension of technological mediation. A new generation of mobile devices are placed side by side with traditional media and the generation of desktop computers in outlining and forming the daily experience of life: the whole of existence itself is immersed in hyperreality.
>
> Devices, like smartphones, tablets, ultrabooks, ereaders and hybrids are *homo cognitivus*' physical mediation tools with space-time, in a continuum where living bodies, machines, codes, data and networks interact: this is the biohypermedia environment, a term derived from the assembly of bios/biopolitical and hypermedia.
>
> Connected and "wearable" technologies subordinate us to a multisensorial perception where real and virtual spaces blur, extend and amplify emotional stimuli.
>
> How are social relations of production and power changed in this new context? What impact will the next nomadic services "offered" by neurodigital corporations have in capturing our attention and influencing the intentions of the always connected? Are we facing yet another structural transformation in the digital society? Above all, what are the conceivable paths and escape routes for mobilizing the multitudes?[5]

Biohypermedia can be broadly defined as the environment in which the whole body is connected to network devices in such an intimate way as to enter into a symbiosis where reciprocal modifications and simulations take place. While the function assumed by

5 Griziotti, 2013a, 2.

machinery in modernity imitates and reproduces human skills, as we have previously discussed with the example of the assembly line and digitally controlled machines, biohypermedia is where

> the main pressure of micro-electric seduction is, in reality, neural in that it focuses on the interpenetration of human consciousness with the whole electronic network. Current digital and communication technologies externalize and electronically duplicate the human nervous system.[6]

This citation comes from an excerpt that Braidotti dedicates to the "posthuman and becoming-machine" and whose hypothesis looks at the fusion of humans and technology. It is a fusion that creates new relations of power "fused with the modified planetary environment" and that "these new […] radical and joyous […] relations support the vitalist ethics of mutual interdependence between species." This post-anthropocentrism is a new utopia, and we very much need utopias at this moment. For this postanthropocentric hypothesis not to remain merely a star shining light years in the distance, we should begin with the current modalities cognitive capitalism uses to appropriate the use of these very technologies with the goal of recomposing social relations of production to "simulate efficiency and cutthroat opportunism." Taking the centrality of technological mediation in the formation of the subject for granted is not sufficient for moving beyond the current situation. A first step may be, at the very least, determining the sense that can be attributed to such mediation in order to understand that, before being post-anthropocentric, we should try to avoid the unsustainable impasse into which the financial system seems determined to lead us.

Genealogy of biohypermedia devices

As sometimes happens, the fusion of two or more technologies gives life to a new paradigm that goes far beyond the sum of the functionalities present in each of them. As we saw in the previous section, biohypermedia is the interactional and integrational environment of ICT

6 Braidotti, 2014, 97.

and the sphere of life. It is brought about by the combination of two of the biggest and most recent technological revolutions: the internet and mobile communications, described above.

Terminal devices are themselves the heirs of two long, technological lines: computers and cellular telephones, to which we could add the branch of Personal Digital Assistance (PDAs, or Palms). In their genetic patrimony, the technological chromosomes of computers and cellular phones combine: computer architecture, microprocessors and operative systems including Unix BSD, Linux or even Windows and SIM cards, antennas and transmission management chips. The transformation of cellular phones into evermore sophisticated, diversified and connected mobile devices, through internet networks like the web and apps,* lies at the origins of biohypermedia.

Such devices continue to expand in number and to evolve rapidly in volume, typology and form-factor. Smartphones, and hybrid phablets,[7] remain the first and main instance: starting in 2010, we see yet another phase of expansion, just ten years after the rapid spread of cellular voice services. Around 1.5 billion cellphones are sold every year. The smartphone, in all its various forms, is now a universal device along with those that already exist and the integrated or peripheral devices that are, among other things, able to perceive and exchange data from the surrounding environment, understand somatic movements, monitor the internal functioning of a body and interact with sensorial input.

Biohypermedia is a new informational paradigm that breaks and subverts previous frameworks to interact with networks. In the first era of mobile telephones, practically only voice communications were possible, while internet access was only feasible through a PC and a landline modem, often using the same line as the home phone. In the ex-nihilo creation of GSM (or equivalent) digital systems and architectures by new mobile operators, the text messaging functionality was considered secondary but provided anyway, even though nobody believed users would adopt it or find it very useful. No one foresaw the explosion of texting that went so far as to influence the very language and written forms of "digital natives"; nor did anyone envision the necessity and utility of this supplementary channel of instant communication that, in reality, is less cumbersome and intrusive than voice calls.

7 These are devices with screens, halfway between a smartphone and a tablet.

The text thus becomes one medium in the mobile spectrum, just as instant messaging would accompany internet content in the '90s. Initially, one of the first freeware programs (a term that indicates proprietary software that is freely distributed and not to be confused with free software) of this kind, ICQ (to be read as "I seek you"), was adopted by programmers working on the same project from different/distant sites. They needed an easy and instantaneous way to communicate that was less invasive and attention-monopolizing than the telephone.

Following the spread of the first digital mobile networks, for example GPRS in Europe, the base conditions for the birth of biohypermedia were created. At the beginning of the 2000s, a tiny, unknown Canadian telecommunications company, RIM, invented the Blackberry, a system and a terminal, primarily for professionals, which made email nomadic. Devising the "push" function, which would then limitlessly extend to other mobile applications, email became perfectly intrusive: not only did it come directly to a terminal that was always with the user, but it also immediately signaled the arrival of new messages, tirelessly and relentlessly. It is precisely this characteristic, so fitting for the rupture of the line between work and free time, that decreed the immediate success of the Blackberry, first for professional activities and, later, in the private sphere.

At the same time, only Microsoft, in collaboration with several contractors, proposed a device that would be a precursor to toady's smartphones: the Pocket PC (PPC), but it would have a distribution limited to professional use owing to the bulky digital pen and somewhat awkward graphic interface. Microsoft had, in fact, adapted a version of Windows to the PPC without taking into consideration that this graphic system was absolutely inadequate for the physical characteristics of the minute display and the operating conditions imposed by mobile use. In addition, PPC software was limited and unstable due to scarce investment.

It seems that it was Bill Gates himself who didn't believe in the future of mobile internet and who, despite being solicited by his close collaborators as well as external consultants to improve the calamitous program for synchronizing data between a PC and the Pocket PC (ActiveSync), he refused to invest sums that were, in respect to what was at stake, risible. In any case, Microsoft's management absolutely didn't consider the hypothesis that networks and mobile technology were a new paradigm that required calling Windows itself into question. The corporation's rigid proprietary and pro-copyright approach did the rest. This attitude would cost them their leadership position and billions of dollars.

Steve Jobs and Apple, instead, understood how to best exploit free software and open source because they had already experimented with rewriting Unix BSD into the operating system Mac OSX. Having intuited the potentialities of a mobile device that would allow for ad hoc applications to interact with networks, he used the same approach to assemble a technological object that was aesthetically and ergonomically conceived for mobile internet: the iPhone was launched in 2007.

Google followed suit but, with their mobile operating system Android, derived from Linux, adopted an approach that was almost diametrically opposed. The objective wasn't, as it was for Apple, to make money by selling technology, hardware and "closed" proprietary software, but above all to occupy the same position in biohypermedia that they had gained in the internet with their famous search engine and to earn profits in the same way. For this reason, Android is open source, its license is free (not including its various accessories) and has been adopted by the majority of producers, with the exception of Apple. Beyond the politically significant fact that these systems are based on common production in free and open source software, these platforms introduced a paradigmatic shift.

Settlers of the New Frontier

The myth of the New Frontier is often invoked in reference to cyberspace. Capitalist colonization in the era of biohypermedia profoundly differs from the venture into the unknown territories of the Far West, where indigenous populations were massacred in order to impose WASP "civilization." The biohypermediatic sphere is a universe that, unlike the big bang, densifies while expanding. There is no sense in speaking about limits in cyberspace. On the contrary, while physical space is limited, cyberspace continues to generate itself and create new resources with a variety of infinite dynamics. Numerous other great media revolutions have shaped civilization, restructuring the human mind and, consequently, changing our very consciousness. Marshal McLuhan developed this intuition, describing the "making of typographic man[8] "from which we can suppose that, while typography obliged us to rationalize the process of reading, more contemporary forms of media more directly involve other senses, such as hearing and sight: cinema, radio and television. Never before, however, despite the "Global Village" foreseen by McLuhan, have we found ourselves so immersed in a media-universe like that of biohypermedia.

Marketers and advertisers talk about channels and multichannel strategies when one of their "campaigns" utilizes several distinct mediums, including mobile internet. This vision may already be surpassed because, in the era of biohypermedia, the surrounding reality is so profoundly innervated by networks that separation into individual channels becomes reductive. One of the parameters that provoked the flop of Facebook's initial public offering (IPO), was probably derived from this error. Many financial operators, on this occasion, blamed Zuckerberg for not having a "profitable" project for expansion into mobile devices. This eventually pushed Facebook to change strategies and to fully exploit biohypermedia. In 2015, around 75% of total advertising profits on Facebook – corresponding to about $15 million dollars – came from the use of mobile devices like tablets and smartphones.

Market laissez-faire in the era of the cognitive nomad consists in favoring the implementation of mobile apps that allow every residual space left in our lives to be reached. The biohypermediatic sphere includes all existing media but is not their simple sum from the moment in which the material of space-time is overlapped and integrated into networks because it becomes the natural environment for biopolitics. While, in the colonial era, imperialist expansion was characterized by the presence of an exterior, an outside, made of territories to be conquered and made accessible for the exchange and exploitation of raw materials to be later integrated into capitalist production[9]; in biohypermediatic space-time, there is no outside to mention, but one or more dimensions that are integrated into material reality and that transform it. First of all, this space-time differs from the space of the new frontier as we know it. Even if we are far from the conquests of the new world, it isn't entirely clear if this transfiguration is created by the overlap and, in certain cases, the fusion or substitution (as is

8 McLuhan, 1991.

9 See Negri, Hardt, 2002, op. cit., 211.

the case with Virtual Reality*) of reticular dimensions to the extent that we could speak of a sort of exodus; this at least if we don't take advantage of these reticular dimensions in order to shift its meaning to intend a post-capitalist mode of being.

Bring your own device[10]

Increasing mobile internet traffic[11], at a rate of around 12% annually, confirms the coming of biohypermedia, the environment in which bodies, lives, machines, networks, codes, data, fluxes, territories and time interact. Mobile devices, beginning with smartphones, are *homo cognitivus*' physical mediation tools with this new environment. Like other components of ICT that come with increasing pulsations, these devices are continually mutating, hybridizing and sprout articulation and ever new technosensorial synapses.

In this new context, relations are changed, including productive relations, power relations and interpersonal relations. This is yet another structural transformation that preludes an evolution of the digital society.[12] Marketing and communication campaigns portray these devices as commodity fetishes, characterizing a new phase of a cognitive capitalism that continues its metamorphoses. The same thing can be said of the automobile in the second part of the industrial era and then again with personal computers. Finance, the impetuous sponsor obsessed with competition, drives the success of the digital big four, the notorious GAFA[13], that have, today, taken the place of General Motors and Exxon in the rankings of global capital. Not even Steve Jobs, in his maniacal search for Pure Form to stimulate in his clients the feeling of belonging to an elite, could have envisaged his iPhone becoming a sophisticated and widespread key to the new ecosystem of biohypermedia.

---

10 Parts of this section were previously published in "Under the regime of precarity: bring your own device," https://www.opendemocracy.net/giorgio-griziotti/under-regime-of-precarity-bring-your-own-device

11 See the evolution on the website http://gs.statcounter.com/#desktop+mobile-comparison- ww-daily-20081201-20150209.

12 "A new society emerges when and if a structural transformation can be observed in the relationships of production, in the relationships of power, and in the relationships of experience," Castells M., 1998, 340.

13 Acronym for Google, Apple, Facebook, Amazon.

The use of these devices, closer to the mobility of a car than the sedentary PC, imposes a new paradigm of the app*. Hundreds of thousands of apps, made available in just a few years, originate from the intersection of two factors:

- The existence of a diffused General Intellect that, forged in the cauldron of the common production of software, has the operative skills to develop apps for mobile platforms founded on free software "privatized" by Apple, Google, etc;
- The desire and necessity to have simple, functional and fast devices able to do specific tasks on the go.

The phenomenon of apps was first instigated and institutionalized in the AppStore*, a special application that is an enclosure where Apple authorizes (or not) third-party apps while, in the meantime, extracting rent from the labor of software developing communities.[14] Apple likes to play on ambiguity in a form of propaganda that exalts the "revolutionary" spirit of technological innovation as a way of consolidating their strictly neoliberal economic policy. Benefits for shareholders, managers and software architects, are coupled with rank exploitation of "low level" employees and subcontractors, like the young precarious workers who run Apple Stores and the semi-enslaved Chinese workers who frenetically build iPhones and other devices in the city-factories of Foxconn.

It is clear that the first iPhone, in 2007, inaugurated the new phase of the cognitive era, just as the PC heralded the decline of the industrial phase in terms of political and financial importance. However, in just a few years, though Apple has lost its monopoly, smartphones and tablets are sold at a rate incomparable to that of the PC.[15] To better trace the cartography where common production and rent extraction intersect, it is necessary to politically analyze the role that devices can assume, also through analogies and differences with other key devices in previous eras, like the automobile and the PC. More and more, antagonistic roles are created that go from a tool of the common in the

---

14 Apple receives 30% of everything sold on the App Store. In 2015 this resulted in twenty billion dollars.

15 Less than five years passed before one million were installed, compared to the thirty years it took for the PC.

hands of an emerging networked movement to bait for new forms of the pervasive exploitation of diffused precarity. Thus we find a continual tension between the living labor of the multitude and the attempt to trap it in immaterial and invisible fences through coercive procedures that transform it into dead labor.

The automobile, too, is a tool for crossing territories, but its use value is hardwired into the central functions of transport and travel that, during the industrial era, were at the center of productive dynamics and life. In the previous section we analyzed how, in the '80s, the arrival of the PC helped break the mold of industrial innovation with its subsequent diffusion of these new flexible and affordable computers throughout what was to become the knowledge-based economy subsumed by cognitive capitalism.* Now, a few decades later, the relation between the affirmation of mass intellectuality and these sufficiently mobile, flexible, reconfigurable and affordable machines is clear.

As we have previously examined, these devices integrate, bolster and make older technologies portable. The key change lies in the combination of miniaturizing and mobilizing these pieces of equipment, allowing them to be always within the body's reach at any given moment, both inside and outside. In an era dominated by emotions, the constant exchange of our senses with the network is central and these small and powerful machines have become a personal *longa manus* for remote actions. The mobile device and its increasing amount of peripheral sensors interact with sound, sight, touch and words, allowing geolocalization, long-distance control and exchange with communicating objects: they augment physical reality* with every kind of information – commercial, cultural, ecological – and act more and more as biomedical sets for the control, correction and support of our vital biological functions. Not to mention the more anthropomorphic peripherals that will complement or replace screens.

We'll later take a closer look at the vital implications that a territory evermore populated with connected objects, denominated with the now trendy term the "internet of things."

While the PC has a central function as the medium for linguistic and written processes and its typical uses often tied to the logical concatenation of thought and the priority management of sequences over time, mobile devices concentrate complex perception of stimuli in terms of spatial organization and intervene in emotional responses. PC work of this kind occupies our left hemisphere, while the continual interactions of a smartphone engage the right hemisphere.

Contrary to automobiles or, to a lesser degree, PCs, use value in network devices is no longer only determined by the characteristics laid out during conception of industrialization, but can radically change in the hand of the cognitive user. Once the internal barriers – created to bridle its use – are broken[16], the user introduces value through settings, updates, multimedia creations, data enrichment and downloading music, games and applications based on their dynamic uses in life and work and based on the users' feelings and desires.

Although the material characteristics remain, when the device is shaped through use over time, its capacities and its performance evolve to the point of no longer having even a remote relation to its initial state. The living labor of the single individual enters into relation with the network to give life to a continuous transformation that generates an abundance that represents both value as well as a common force that continually breaks the limits weaved by biopower. These activities don't happen in an exclusive human-machine relation but through common-based peer production, the social production of the common on networks that are not just lining up to endow capitalism with yet another socially acceptable era of exploitation. The innumerable websites, blogs and forums in every language globally blooming and growing are a workshop for worldwide exchange where the digital means are developed that could be used to free workers both from precarity, from total subsumption to a life of labor and to unhinge the enclosures of digital corporations. With Apple leading the charge, the latter attempt to limit the escape from their enclosures in order to better exploit both individual life and common-based peer production.

However, large technology companies encounter new and unforeseen resistance in their systematic search for the planned obsolescence of their product[17] in order to boost consumerism: in Paris, London,

16 Jailbreaking is a process that allows the "prison" of the App Store to be broken and install applications on Apple products through alternative distributions. Rooting, on the other hand, is the same function on Android devices and the Play Store.

17 Apple, for example, conceived of an iPhone that can cost up to $1000 US dollars with a non-substitutable battery. The life of the phone is, in principle, limited to that of its battery. Collective ingenuity has found a solution: low-cost kits are available online, as well as the stores mentioned above.

Milan or Shenzhen,[18] we find entire alleyways of Asian workshops worthy of a scene from Blade Runner;[19] for just a few dollars, you can repair any device or prolong your smartphone's life by changing the built-in battery or replacing a broken screen in under an hour.

Even when capitalist control isn't able to turn a direct profit from common production, it still tries to create new situations to capture it. Until recently, large companies' employees were often supplied with a laptop and a mobile phone with a subscription. Despite hesitations related to "security" problems, today companies allow or even oblige people to use their own devices: BYOD is transformed into an imperative. This not only to apply a deep cost-cutting policy despite generous bonuses to managers, traders and shareholders but, mainly, to exploit our productivity through machines that we have ourselves configured.

Switching and the aristocracy of cognitive startups[20]

Many critics, including critics on the left, express deep doubts about the capacity for bodies to interact with the network. Interactions are said to be hindered by the technical characteristics of the terminals, for example, the low audio quality of an mp3 or the scarce graphic resolution on smartphones and tablets.[21] However, these arguments have now become irrelevant in that, as one can imagine, evolutions are continuous and, even among the most common smartphones, there are models that have successfully substituted entire ranges of compact cameras. Even more significant is the progress made in sensorial qualities of numerous accessories, gadgets and probes that have increasingly expanded capacities that can sometimes surpass biological human senses. To these possibilities we can add those coming from new technical evolutions like augmented reality and dynamic hologram management.[22]

---

18 The Chinese Shenzhen Special Economic Zone is near Hong Kong where there are many device factories: the population rose from 20,000 to 13.5 million over the past 30 years.

19 Scott, 1982.

20 A precarious/temporary worker doing intellectual work.

21 Ippolita, 2012, 73.

22 A fine-tuned version of the animated hologram stunned the first viewers of *Star*

The failure of the Google Glass project as a new gadget to mass produce cooled the enthusiasm for the hypothesis of an internet that is more and more present in our lives and on our bodies, but it didn't stop the general movement in that direction. Besides, Google's wearable network glasses will find other outlets in diverse "professional" uses, from surgical training to law enforcement applications. The application potential to gather, elaborate and consult information through biohypermediatic devices is still in the early stages of development and is only now coming into a phase where multimedia switching is starting to play an important role. Multimedia switching is the possibility and the ability to send an information flux through thousands of levels in a physical and network reality where humans, connected objects and automatisms interact. Manuel Castells spoke of this phenomenon, affirming that at the heart of the power of News Corp, Rupert Murdoch's media empire, we find switching power, which he defines as "the ability to control connection points between different networks (e.g. business, media and economic networks) [as] a critical source of power in contemporary society."[23]

The same principle can be extended to the level of individuals or any collective instance in the sense that switching is a nodal point at the center of the knowledge economy. For the majority of the digital generation, abilities relative to switching are acquired very early and indissolubly integrate with other skills assimilated during periods of learning. The most widespread and banal example is the continual search for "viral" information, a behavior born as offering to promote something produced by users for users thanks to social networks. This switching capacity is a complementary skill, similar to the driver's license for the previous generation, but much broader, evolved and expensive, along the same lines of the diverse complexity of the two types of technology, the automobile and ICT devices. In switching, creative ability resides therefore at the base of the skills required of workers in the knowledge economy and is also a junction in the dialectic between autonomy and control, both in life and in creative work. Picking back up on the analysis developed by Ursula Huws,[24] we can consider the ability to manoeuver through biohypermedia as

---

Wars in 1977.

23 Arsenault & Castells, 2008, 488.

24 Huws, 2010.

an indispensable skill in the "dialectic dance" between tendencies and countercurrents. The direction of public or private entities cannot do without these skills, but must also maintain a difficult balance between the insatiable need for a continuous flow of innovative ideas and the equally imperative requirement of maintaining control over intellectual property and the management of creative forces. For their part, cognitive workers search for an equilibrium between the necessity to express their creativity and the obligation to earn a living.

In this complex situation, made of contradictory relations, collusions and conflicts between the main actors, do we not risk the emergence of an cognitarian startup aristocracy? Contrary to the workers' aristocracy in Leninist memory,[25] these "aristocrats" are not the elite factory worker but rather indentured servants in startup companies. Above all, they are put into incubators or in tent camps, conditions considered "privileged" for starting in but that are, in realty, not always very comfortable. In the case of tent camps, "partners" of a few startups meet for a few months squeezed into large open spaces, where each team has at most one table, a few chairs and a Wi-Fi password.

The desire to kick off a new project and the enthusiasm that accompanies it are positive impulses and being a part of a startup can be an educational and important experience, but the way in which these positive stimuli are manipulated and channeled is often reprehensible. There is something strange in the infatuation that floats around these movements and the mainstream media that glorifies them with stories that sprout from the well-fed American mythology of Bill Gates and Steve Jobs' garages. Institutional agencies are created to incentivize the development of startups which, in itself, wouldn't necessarily be a bad thing, especially in Europe where, in countries like France, Italy or Spain, youth unemployment rate is already high and continuing to grow. More worrisome is the startup infatuation with a system that allows young people to be removed from unemployment lists at their own detriment. Often, they find themselves in a condition of self-exploitation and self-precarity that is even worse than that of their colleagues with contacts and, what's more, take on considerable personal financial risks. The so-called freedom of not having a direct hierarchy to answer to is substituted with the control and stress imposed by banks, clients, paying taxes and the duties and responsibilities tied

25 On this topic, see *The Short Twentieth Century* (Hobsbawm E., 1994).

to administrative fiscal procedures. In addition, there are no limits in working hours in a startup and very few make it, so the result for the great majority is that of a starving wage. All this without counting the fact that, beyond very few exceptions, startups are particularly fragile and generally unprotected from the often-predatory practices of corporations and public entities (late payments, blackmail, extortion, etc.).

Video games and the art of escape

In apparent contradiction with the voluntary forces of startups who have little free time, video games have become a phenomenon of historical proportion and are at the heart of a sector in which the annual revenue is nearly double that of international cinematographic production. These are statistics that have a symbolic value and seem to indicate a global leadership position for video games in the domain of entertainment. Some even maintain that video games are an invention as important as the printing press!

Our analysis begins with the maybe self-evident affirmation that, in the term "video game," the most important part is "game," because the playful aspect is the most prevalent. When the scope of the game changes, even if the technology is the same (like in the cases of professional training and in extremely dramatic operative situations like piloting war drones), the user obviously does not fall into the same order as a normal gamer. It might then be important to make a first distinction between the three main typologies[26] of users and video games and, somewhat arbitrarily, take above all the third category into consideration.

The first category are the casual gamers, those we encounter most often in public transportation while they play with their smartphones. These gamers prefer simple, scored "arcade"[27] games which are often

26 In addition to these three typologies, a fourth could be added: those of professionally paid players. A small amount of players who turn professional participate in video game competitions and tournaments, are sponsored, win large prizes and are comparable to the commercialization of professional sports.

27 For example: *Space Invaders*, *Defender*, *Asteroids*, *Tetris* and *Pac Man*. These are smartphone or tablet versions of the first arcade video games that are played in public places with coins or tokens, physically constituted by a machine in a box.

smartphone versions of first generation games like Tetris or Pac Man. This activity probably substitutes others like, for example, reading a newspaper or a book, or it can replace behaviors like the "catatonic stare" or "sleeping standing up" and serves above all for relaxation or reducing the stress of having to spend long periods in unpleasant conditions to go to work in a metropolis that is increasingly crowded and polluted.

The second category comprises family video games. Here, too, we see activities that are not dissimilar to those that were once done with board games – activities that entail a physical sociality among friends or family. In this category, preferred games are team-oriented and sports-like, for example, soccer and the Nintendo catalogue, including the famous Mario Bros series.

Let us instead ponder over the third category, the pro gamer or so-called hardcore gamer, i.e. the category of the most tenacious users who dedicate a substantial amount of their time to video games. Their preferred tools are consoles (PlayStation or Xbox) or gaming PCs. Users of gaming PCs, other than being passionate about video games are often also partially hackers who assemble their PCs and are able to "crack" and modify games. Hardcore gamers have a wide variety of games at their disposition, among which stand out adventure games like *Assassin's Creed*, action games like the notorious *Grand Theft Auto* (GTA), first-person shooters like *Battlefield*, and strategy and war games not dissimilar to the classic, chess. A whole other discussion would have to be reserved for role-playing games, known as MMORPG (*Massively Multiplayer Online Role-Playing Game*) that entail the simultaneous online participation of many gamers. Thousands of players can interact in synchronicity, assuming the roles of characters who live and evolve in a context that constitutes a permanent and coherent world. Here we are facing a phenomenon within a phenomenon that has to do with a portion of hard-core gamers who use the PC to such an extent that socio-psychological studies have been conducted to understand the motivations that push these participants to spend, on average, 22 hours per week in full immersion.

Without, however, concerning ourselves with the details of such analyses, here we will limit ourselves to finding a common denominator for positioning video games in the complex puzzle of contemporary technological mediation. From an organizational point of view, what categorizes the video game industry in the pro gamer category is the frenetic activation of an army of scriptwriters, graphic designers, software coders and editors as well as console or component (graphics card) manufacturers for the creation of virtual worlds that are more and more immersive, dynamic and captivating; all of which allows them to completely distinguish video games from the passive time spent in front of a screen (TV, streaming services on PCs and tablets, etc.) while watching a fictional work (films, series, etc.).

This race to a parallel virtual reality seems to respond to the motivations expressed by players[28]: the need to escape a world that they find unpleasant, stressful, coercive and a life hungering for recognition. The pseudo-reality of the game allows them to assume other identities through avatars and, at the same time, be themselves. An environment open to all that lets them accomplish "impossible" feats in scenarios endowed with an impressive realism. An environment that permits them to be appreciated and obtain public recognition in a world that is plagued by unemployment and precarity to the point that many young people can only swell the ranks of the neets,* having renounced all hope.

Video game producers, creators and writers aim to obtain higher and higher levels of psychological realism, a realism that players experience through the interaction with these virtual worlds, coherent with the evolving dynamics in respect to a world of falling prospects and expectations. At the same time, through the quality and fluidity of the virtual environment, video game creators tend to create an immersion phenomenon that directly involves the gamer's senses, isolating them from real stimuli and entailing an intense use of cognitive resources in the game's context. This last aspect is a first explanation for the apparently surprising fact that their levels of concentration are actually higher during play time rather than in their deluding and precarious "real life."

Certain games that require attention to be focused exclusively on a single objective also seem to induce a sort of trance characterized by an insensitivity to external stimuli and an attenuation of consciousness not unlike that provoked through the powerful sounds of techno clubs. The next step on which companies like Facebook are concentrating notable resources is Virtual Reality (VR). VR will be a leap in the direction of further immersion and presence that extends sensorial

28 Yee, 2007, 773-775.

implications and isolation from the real through peripheral devices – for example cardboards, inexpensive visors in which a smartphone is inserted – that act upon the senses in a wider and more involving way.

Regarding apps, a new current of even more realist video games is emerging. We are dealing with veritable films[29], shot with famous actors but that remain interactive. The gamer can embody one of the protagonists of the film and, based on the choices made during the game, can change the plot of the story. Taken together, these developments allow us to imagine a new paradigm of total immersion, of which we still don't fully know the psychophysical consequences.

The video games of pro gamers already grant the possibility to live out a desire for escape from the real and an immersion into the virtual – an apparently "easy" escape with a ludic and gratifying component at their fingertips. This ease seems to contrast the difficulties of living in the "real," a diffused form of renouncing the uniformity of a system of commercialization and the feeling of helplessness that prevents any reform or radical change. Taking refuge in virtual worlds is only in apparent contradiction with the movement of autonomous cooperation that we will later address because it is still a form of research and the construction of a different reality. In both cases, there is a sort of pursuit for an "outside" with the sole difference that taking refuge in a video game represents a form of simple and reversible escape.

At this point, we could ask a few "politically incorrect" but pertinent questions regarding the role of video games in a society with Unconditional Basic Income (UBI*). In the perspective where basic income is guaranteed, wouldn't video games risk doing the lion's share for the undereducated sector of the population or for people who are less motivated to better themselves through work? Don't video games represent a deviance that harkens back to Marcel Duchamp's "bachelor machines," referred to by Deleuze and Guattari as "recording surfaces, bodies without organs" that establish social schizophrenia? Perfect and attentive machines that are an end in themselves. Isn't there also a "relation with ludic technology inclined to pleasure that is not based on functionalism[30]"? Or will the possibility of changing

29 See for example *Beyond: Two Souls*, directed by David Cage, 2013, with the participation of actor Willem Dafoe.

30 Braidotti, 2014, 9.

the real without an escape into the purely virtual give video games a different function? What about the scenario of *Her* (Jonze, 2013), or the birth of an intimate relationship between a man and an operating system bestowed with artificial intelligence?

Robot intelligence and unconditional basic income

While biohypermedia represents the extension of the internet into bodies, emotions, affects and feelings that are not necessarily mediated by writing or even language at all, the internet of things is another extension to the rest of the physical world. The considerations on the reterritorialization generated by this technological mediation as a doubling of the world implies an enormously rich field of thought that is only today starting to be seriously explored, like in this interview with the French philosopher Bernard Stiegler:

> The internet of things represents the extension of the internet to objects and places in the physical world. It is in this context that generalized automation is produced. There is no longer any need for cashiers, drivers, workers, handlers – and, in the end, of anyone other than man-for-man services for all those who cannot afford a robot. Everything will be automatized. [...] The extension of [...] IPv6 (an abbreviation for the sixth version of internet protocol) [...] generates automatic "intelligence" of which the gains in productivity resulting from this new automatism will be gains in time that will pose an enormous problem constituting a great historical alternative.[31]

Stiegler anticipates a world of generalized automatization. First off, we can see this as an extension of industrial robotization to the world and life at the precise moment that Foxconn – the global factory par excellence with 1.2 million workers and a 40% share of consumer electronics production – has just deployed their first army of thousands of their own robots, the Foxbot, for the assembly and testing of the iPhone 6.

---

31 Stiegler, 2013.

Collective human intelligence is, and will increasingly be, faced with a distributed robotic intelligence that generates free time and unemployment. No recent social model seems compatible with this prospective of generalized automatization and precarity/unemployment. Keynesian Fordism and, in general, the consumer model are in contradiction with this hypothesis of robotic manufacturing. Who is going to be able to buy the merchandise and products produced by robots? Cognitive capitalism, where this botsociety is developed, therefore seems to be moving towards a socio-economic and political impasse.

A possible uncontrollable systemic crash would entail a serious risk of regression and for which certain ominous signals are already visible in Europe, like the danger of racist, obscurantist or fundamentalist social movements or regimes taking power. The latter, by definition, are oriented toward exclusive, frontal and civil conflict against one another with violence of fascist inspiration. This is not recent history repeating itself, if not in the sense of a Great Crisis that could give rise to a generalized and destructive catharsis or to widespread but fragmentary and discontinuous civil war.

There are paths to avoid further tragedy. First and foremost, unconditional basic income (UBI)* should be considered as primary income available to every adult citizen, given the impossibility of returning to the "full employment" of the industrial era. Moreover, extensive economic studies show that financing UBI is possible.[32] Vercellone presented the argument for UBI in a recent interview:

> "Minimum income," which unconditional basic income (UBI) could represent, would immediately favor the passage from a model of precariousness to a model of chosen mobility, other than the liberalization of the living forces in an economy founded on knowledge.
>
> UBI would correspond to a form of flexible reduction in working hours extended across the whole of life and presents, in respect to the reduction in classic working hours, an additional advantage: that of reinforcing the negotiation power of the work force. In effect, UBI would change the power relation inside

32 Monnier & Vercellone, 2013.

> companies. The employer would reduce the amount of precarious/temporary labor in order to keep workers that, in any case, have an alternative. In particular, a shortage of manpower would be created in the economy of industrialized services (like McDonald's) that today consumes a great quantity of precarious labor. As a result, even in these sectors, there would be a dynamic that favors a depart from Taylorism.
>
> UBI would equally favor negotiation power for other unsalaried employment categories that are evermore common. Thus, for example, the self-employed or what in Italy are called "second-generation autonomous workers," for whom a departure from official salaries is only a formality, could benefit from wider margins for maneuvering through subcontracting and contractual relations. In effect, UBI would allow them to reduce working hours without suffering from decreases in income.
>
> Furthermore, the association between the permanent guarantee of income and the reduction of working hours, granted by UBI, would favor the transfer of manpower from sectors oriented toward the logic of commercial profitability to the non-commercial sectors in the social and fair trade economy, as well as the commons of knowledge.[33]

A few thinkers, including Stiegler, believe that it would also be beneficial to add to this untouchable basic income incentives to participate in projects of the common,[34] creating a social relation of production that is not exclusively dominated by financial rationality.

---

33 Vercellone, 2015.

34 "The common is therefore the product of a social and institutional construction that elects it to this status. It refers not to an essence that precedes it but to the forms of governing and labor cooperation that assure its production, reproduction and distribution. As such, the common potentially concerns every type of resource, good or service, even if this does not mean overlooking the particular problems of management that each of them would present." Vercellone et. al., Managing the commons in the knowledge economy, 2015, 52. See the Glossary.

Once again, we find ourselves facing a deep political choice that would avoid getting to the point of a machine-guided determinism. We are facing two dangers: the first is that of preventing this diffused automated intelligence from being developed chiefly in the current vision of economic rationalization that favors social injustice and the deepening divide between the oligarchy and the multitudes. Things aren't as simple as they seem and the current system hasn't yet played its last ace. The ten-thousand Foxbots mentioned above assemble devices that are sold at $800-$1000 US dollars to "privileged" consumers. A great number of Chinese manufacturers[35], however, are able to furnish devices with Android's fork,[36] which are still quite sophisticated and at much more accessible prices – even for the Chinese workers who assemble them – and this seems to be a repetition of the phenomenon of standardization of automobiles sold to factory workers in the Keynesian-industrial era. The difference lies in the fact that, over the long term, a good part of the workers will be substituted by robots and will not be able to continue to consume without some UBI-like mechanism. Obviously, history teaches us that the solution isn't found in sabotaging or destroying robots.

The second danger is much more abstract and distant. It regards the choices in governance mandates of this diffused automatic intelligence. In practical terms, we are dealing with either operating (or not) political control on the level of trust we give to artificial and algorithmic intelligence as a diffused biopolitical agent. However, this is a subject that is more relevant to organizational methods and will thus be developed later in the third section of this book.

35 Among them we find veritable multinational corporations like Lenovo and Huawey, and many start-ups, in an economy that now extends to other emerging Asian countries such as India and Indonesia.

36 A fork (or branch), in IT and software in general, indicates the development of a new software project starting from the source code of another, pre-existing program.

Electronic waste landfill in Agbogbloshie, a suburb of Accra, Ghana. Photo courtesy Valentino Bellini / The Story Institute

SECOND EVASION

E-Waste 2.0[1]

E-WASTE IS THE NEW EMERGING PATHOLOGY OF THE MAN-ENVIRONMENT ecosystem, born from the current phase of capitalist production. Electrical and electronic waste (e-waste) is today the fastest growing source of waste in the world. Every year, between 40 and 50 million tons are generated and, according to a study by the UNEP (United Nations Environment Program), this figure could grow by around 500% over the next ten years, above all in countries like India, China and a few regions in Africa where the technology industry is rapidly developing. It is dangerous waste that contains dozens of toxic substances (both for humans and for the environment), is difficult to process in a sustainable way and requires costly procedures to recycle. This is why around 80% of the e-waste produced in developed countries (with North America and Europe at the top of the list) are not processed on site but are loaded onto cargo ships and sent to developing countries – often illegally – and where – often illegally – they are processed.

In our history of subjecting and exploiting the earth's resources, humans have always generated scraps and waste as the collateral of our manufacturing and consumption activities. But it is only with the

1 Article by Gianluca Giannelli and Giorgio Griziotti (2013), written at the request of the photographer and militant Valentino Bellini for The Bit Rot Project: www.bitrotproject.com.

so-called industrial economy, including the chemical, petroleum and plastic industries, that the waste produced by humans was no longer metabolized and recycled by the natural forces that regulate the vital balance of our planet.

The process of commercialization and capitalist valorization has thus created a veritable "waste economy." This economy aims to extend the logic of profit and exploitation to the scraps that it itself has made in an endless iteration of the process that extracts value even from its own death. Like organic waste that through decomposition is transformed into organic material for regenerating life, capitalist command forces human beings to decompose the waste generated by other human beings in order to create profit. It is no longer worms and enzymes that complete the natural cycle between life and death, producing biochemical energy and fertilizing the earth, but men and women who are forced to decompose inorganic carcasses in order to survive and profit from others.

This strange way of differentiating and specializing functional to the human race, set up by the global organization of capitalist production, first harvests natural resources, knowledge and traditional modes of production (that were once sufficient to sustain local populations) and compels entire populations, now desperate, to accept one specific function in the vast system of the global division of labor in order to survive. Old and new processes of colonialization create old and new specialized functions based on race and territory, thus spawning generations of humans condemned to fulfill specific functions. These are men and women who are "socially modified" to do a precise task that is functional to maintaining the entire system.

This is exactly what has happened for the "waste economy" that, since its birth in industrial capitalism, has undergone changes due to its internal transformations over time. The same goes for these "socially modified" humans, in turn transformed by the shifting functional needs of the economy. We can easily recall the "junk dealer" of industrial cities, or the "metal digger," generally looking for copper, in post-industrial dumpsters. These two figures are, however, the result of a historically and qualitatively different systemic modality. This without mentioning those who sift through solid urban waste, scavenging through the open-air dumps of western industrial cities.

Today it may be easier to imagine – thanks to the information spread by connected communication technologies – men, women and children from African villages who "decompose" large commercial ships that have beached because they were abandoned and left adrift, or the new foragers combing the dumps of Madagascar, a once sublime location now transformed into one of the 21st century's post-industrial global dumpsters.

It is the current phase of cognitive-financial capitalism that defines the physiognomy of our contemporary version of the "waste economy," with e-waste being the primary material and the symbol of discontinuity with the past. It is precisely through the cognitive machines that a new genre of consumer individuality is made; the materiality of consumer objects, despite its symbolic meaning of social emancipation that had characterized it in the Fordist era of mass production and consumption, has exhausted its ability to give pleasure. In order to avoid cognitive machines becoming recyclable tools for the autonomous production of the multitudes, the goal of new millennium capitalism is to create biocognitive individualities destined to generate and consume information, signs and symbols, throughout their biological existence, placing true commercial value in immaterial content.

The cycle of capital valorization that, in the modern industrial era, was represented by the famous formula "M-C-M+".[2] Today, biocognitive-financial capitalism becomes "M-I-M+" where "I" is the information that produces goods through knowledge. This information is continually created and consumed by human beings individualized through processes of segmentation and biocognitive segregation. In order to survive the fall in profits in a now automated industrial production, cognitive capitalism needs a "digitalization of the Ego" for the purpose of continually appeasing its "semiotic bulimia" and from which it extracts most of its value today, leaving materiality to be a mere support, vehicle, a means for the "sign," i.e. the true object of individual pleasure.

This logic of unconditioned commercialization is a machine that produces dichotomies: sign/support, information/material that have repercussions on divergent space-time dimensions. Signs and information penetrate everywhere, all the time through hypermedia networks and individual minds. Supports and material remain limited

2 M-C-M+ with this formula Marx describes advanced capitalism (at his time), where money (M) is used to buy commodities (C) and then sold to make more money (M+). Thus commodity is a means to increase money.

and localized due to a necessary and intended physical technological obsolescence and because of where they are at the end of their use cycle. It is in this procured "spatiality" that we come back to e-waste, destroyer of land and men, born from the neoliberal waste management of digital consumerism. Here, this spatiality meets the concrete, the misery and the cruelty of the living conditions of men, women and children who can only survive by decomposing it. While in the solitude of network life we lose contact with the materiality of technological objects, it is here in this spatiality that we find it in the form of toxic waste. The gazes of the young, adolescent men made to live in it are shocking.

The vastness and quantity of e-waste sites in the poor countries of the south can be explained with the millions of tons of electronic scraps that are poured into them. These flows are only growing because, as we integrate the net into our territories, billions of new mobile devices like smartphones, tablets, laptops, etc. are being added to the already immense amounts of televisions and PCs. This squandering is incentivized by ICT corporations that, in their spasmodic search for infinite profit, implement an increasingly common and aggressive planned obsolescence into their products. Under the regime of financial governance, the laws that should prevent these humanitarian and ecological disasters are written in a way to leave ample spaces to the interests of those who hold public and private economic power. What a difference from the inflexibility with which laws created to keep migrant workers far from our postindustrial paradises are enforced. What likeness to the indifference (or hostility) with which thousands of refugees are left to drown just a few kilometers from Europe's shores. Executive governance largely prefers they stay home – a home they have turned into a toxic dump site.

Strengthening and enforcing these insufficient international laws would thwart enormous profits. In fact, getting rid of an old PC by sending it to an African dump costs only $2 dollars, against the twenty it takes to recycle it in a sustainable way. The $18-dollar difference is divvied up by international "e-waste management" operators and their corresponding local mafia partners. The collusion and complementarity between legal and mafia capital found in certain countries in the south is reproduced on a global scale. In peripheral regions, the forms of accumulation and mafia organization constitute an essential means for the international division of labor. According to economists who are not subservient to mainstream finance: "in the end, cognitive capital and capital linked to the mafia find a true unity in the innate opacity of financial markets where any distinction between the two disappears."[3] This is just another example of how exploitation functions in the same way both in building and in destroying.

On one side, eco-mafias are taking control of rare and non-renewable resources in developing countries, on the other, they contribute to the context of ecological crisis with e-waste. In these two cases, we are facing the expropriation of the common, both in the devastation of the environment and in an exploitation that is borderline slavery due to the harmful conditions in which those who work in such a hell have to live.

The shocking paradox is seeing the tangible results of the crude materialization of the global division of labor right under our noses. Financial oligarchs reap the shares from intellectual property rights, immaterial production and ICT biohypermedia device sales, while postindustrial multitudes are damned to the land of technological dumps that render their ecosystems sterile and toxic. These e-waste territories, the 300 daily tons of radioactive water poured into the sea from Fukushima and the decaying lands devastated by fracking to extract gas and oil from shale are only a few examples: the toxicity of capitalism has no limits. The network, the dominant machine of economic rationality is accelerating the destruction of the biosphere. Until when?

3 Didier Lebert and Carlo Vercellone, "Capitalisme cognitif et capitalisme mafieux," [our translation].

Part Three: ORGANIZING

Evidently, every kind of society can be correlated with a kind of machine: simple or dynamic machines for sovereign societies, energetic machines for those disciplinary, cybernetics and computers for societies of control. But the machines don't explain anything. Instead, collective concatenations which are only a single aspect of them must be analyzed.
—Gilles Deleuze, *Le devenir révolutionnaire et les créations politiques*[1]

Choosing the subtitle "Organizing" for this third and final section merits some explanation. It is a purposefully generic term used to highlight a fundamental problem where technological mediation favors an explosive fragmentation of the present, possibly even more so than in the topics of the previous sections. In our eyes, a reflection on the new modes of organization appears urgent precisely due to the unsustainability of current trends. The use of this gerund does not necessarily imply a specific end or objective, nor is any presupposed ideology intended. Every possibility of considering the early forms of organization that tend towards post-capitalism in the contemporary technological environment is open, including those that cooperation assumes in the context of cognitive capitalism.

1 Deleuze, 1990.

It may be pointless to ask ourselves if the objective to organize labor or life: from what has emerged in various passages in previous chapters, it seems rather difficult to maintain any separation between the two. Consequently, making clear distinctions in the forms of organization based on the context in which they are employed can be a somewhat complicated task. Instead, it seems necessary to analyze the roles that technological mediation plays in the fields of organizational cooperation and in the shifts in social and economic patterns.

We have always attempted to avoid the anathema of "technological determinism" which often afflicts those who deal with technology, without however neglecting to evaluate the grade of autonomy of technological artifacts and the evermore automatic interoperability of their components. We have focused on a few of technology's operative modalities in social cooperation without losing sight of the subjective fabric on which they are exercised. This observation has lead us to begin with the hypothesis that a fundamental change has already taken place: the passage from "belonging" to "crossing," a more adapted term for "nomadism." We are facing an extremely rapid and profound ontological transformation if compared to the time it took for changes of the same magnitude in the past.

In the previous sections, the dynamics of technological mediation over production and life allowed us to at least imagine, if not anticipate, a few of the scenarios towards which we are sliding, shocked by the mechanisms of biopower working against our subjectivity and the emancipative attempts being made thanks to the spread of knowledge. Here, our vision of this panorama is much less clear. Digital technologies have favored the development of dense and widespread reticular horizontal cooperation that is, however, often superficial, ephemeral and confused. In many cases, even if such cooperation springs from autonomous initiatives, they are sucked into the neoliberal tornado, while others are the initiatives born directly from companies in digital capitalism, like Facebook.

This ability to articulate and integrate non-profit initiatives into the capitalist system is not new. Think of the fundamental drift towards capitalism in the world of NGOs and NPOs or the earlier cooperative movement, born in the heroic era of the first industrial revolution. In this context, starting from the second half of the 19th century, financial resources and means of production and consumption were shared "in the name of the right to life" and on the basis of a conception of needs that largely remains outside of mercantile logic and capitalist production. Worker cooperatives are ways of organizing that are non-profit as well as forms of self-defense for those who faced the social and human conditions of that time, born inside the labor movement. In that period, mutual aid societies sprang up and developed, heralding the social gains obtained nearly a century later. For some time now, the cooperative branch that still remains is less and less characterized by a politically defined identity and are often reduced to a form of business with mutualistic ends. This doesn't prevent cooperatives from becoming large corporations[2] that, beyond their legal statute and mutualistic formalities, are no different from normal companies in terms of the treatment and exploitation of their employees, for example.

With such a lack of clarity, the constituent dynamics of post-capitalist organization are difficult to perceive, to the point where internal fringes of antagonistic movements sometimes seem more attracted to the perspective of a simple rupture with the present than to a slow and contrasting construction of consensus on new organizational forms of the common.

It therefore seems necessary to deepen our analysis of the numerous initiatives of cooperation that attempt not only to escape or limit compromises made with the current economic system, but also to create grounded alternatives; a space-time of the possible forms of organization where the modes of collective functioning neutralize the processes of cognitive capitalism's technological control. Before examining these new forms of organization, it is worthwhile to clear the air of any ambiguities regarding so-called ethical capitalism. In Davos, in 2013, Nobel peace prize winners, representatives of NGOs, multinational agro-alimentary companies and central banks[3] debated the "ethical" hypothesis, observing that, over the last two centuries, the capitalist system has supposedly reduced the amount of impoverished from 85% to 16% of the world population and that life

2 In Italy, the foremost example is Coop Italia, a leader in the sector of distribution. In France, the bank Credit Agricole, with over 150,000 employees, is a cooperative.

3 See: S. Bergman, "Il Capitalismo etico? Vale la pena provare," *Huffingtonpost*, 2014 http://www.huffingtonpost.it/stanley-m-bergman/il-capitalismo-etico-vale- la-pena-provare_b_4687951.html.

expectancy has purportedly risen from 35 to over 60 years of age, except in post-soviet Russia.[4]

Even if we believe these statistics to be exact, which in itself is difficult since only those who have an income inferior to one dollar per day are considered impoverished, we could also contest that it is not to the merit of capitalist development that the human condition has improved. It is, instead, a complex dynamic in which, without the struggles of antagonist forces inside the system, the situation and the statistics would not be the same. Maybe we would still be in Zola's time when children worked in mines rather than in Bangladesh's dangerous Benetton, Zara and H&M factories.[5] The inversion of the tendency between income from labor and capitalist rent that began nearly 40 years ago is actually pushing towards a phase that is exactly the opposite of that described in Davos. In most economically "developed" countries, the number of poor, unemployed or precarious workers continues to increase and the arrival and permanence of the Great Recession only aggravates this situation.

The society of automatic control

First, Foucault at the root and, later, Deleuze more explicitly in his famous "Postscript on the societies of control"[6] and lastly Hardt and Negri in *Empire* speak about the passage from the disciplinary society to the society of control. The latter write: "the mechanisms of control become [...] evermore immanent to the social field, diffused in the brain and the body. This control is extended much further than the structured sites of institutions through flexible, modularly and fluctuating networks."[7]

While the arguments tied to the modes of autonomous cooperation and those that regard common goods and the common are widened, cognitive capitalism favors the development of a society that is increasingly governed by automatic or cybernetic mechanisms of

4 Mackey & Sisodia, 2013.

5 See the fall of Rana Plaza, a commercial building in the suburbs of Dacca where 1129 people, mostly workers at the textiles mill there, died. http://www.repubblica.it/esteri/2013/05/08/ news/bangladesh_crollo_palazzo_rana_plaza-58320497.

6 Deleuze, 1990.

7 Hardt & Negri, *Empire*, 2002, 39.

control, constituted by algorithms and machines managed by biopower. This was also revealed by Mario Draghi's (voluntary?) slip on the automatic pilot (of financial control) that governs Italy and, what's more, a good portion of Europe, neutralizing any electoral outcome. The interest of capitalist governance in "innovation" thinly veils the use of *téchne* in order to articulate modes of value capture and integrated control over life. The importance of automatic control of behaviors, bodies and social relations is evident. The excess produced by collective intelligence through technologies, of which the multitudes of hackers are an incarnation, are so widespread that only a diffused system of intelligent automatisms can attempt to contain them, exploit them or, as a last resort, repress them.

According to an approximate calculation, less than 10% of the production capacities in ICT are managed and/or remunerated in the capitalist system.[8] This reduced techno-scientific army functions in large part by expropriating and exploiting the autonomous creation of the other 90%. A classic example is that of free software. Multiple activities are put to work in order to reach this objective: from the development of program interfaces to the "affective" osmosis with the world of hackers, all the way to the extension of copyright. On this basis, the dispositives of control are conceived, implemented and embedded in every nook and cranny of the social body and the machines able to transform life into code and code into life.

The automatic control of dynamic systems has, as an end, behavioral modification through the manipulation of entrance points. In the case of highly complex systems, like those relating to life, the difficulty lies in identifying the mathematical models that correspond to the behaviors that are to be modified. Google and Facebook are some of the examples where this strategy has proven successful: both have constructed, within a few short years, a power that projected them into the top group of global capital for their conception of algorithms and software that morph human behavior.

However, despite this intrusiveness, modes of unmeasurable,

8 According to a study by the IDC Institute, there are about 40 million professional software and digital specialists who represent less than 10% of the total number of people with technological ICT abilities. In addition, a rapid progression in this dynamic is underway with the coming of adult age of digital natives. See: http://www.infoq.com/news/2014/01/IDC-software-developers.

unpredictable and autonomous individual or collective action remain. They are so present in the lives of the multitude that they can often be synthesized in a simple qualifying prefix. From horizontal self-communication to self-production and the widespread forms of self-education and self-reliance propagates molecularly throughout the bosom of cognitive capitalism, modifying the forms of cooperation and sociality. The very model of peer-to-peer (P2P) itself is an applicative architecture of network autonomy with a theoretically unlimited potential, to the point that no absolute control in this field is possible. Thanks to P2P, numerous and important battles over copyright have been won, like in the case of more direct access to music and other artistic creations.

The sharing economy remains an open domain where even capitalism itself knows how to use the P2P model to its own advantage, positioning itself as the referee of transactions between peers and raking in a conspicuous rent from the exchanges made on eBay, Airbnb and every new application. In this field, it is possible to innovate in the economic and political sense as indicated, for example, by the experiments with Integrated Cooperatives, but even the sharing economy is largely exploited and controlled. This makes it ever more complex and urgent to examine the dis/equilibrium between controlled and controlling systems. The question that arises is: what are the techniques, the tools and the procedures at the center of this dynamic?

Previously, we have seen how even the limited use of the first Pocket PCs[9], in the period when media and digital corporations entertained the lucrative scam of the Millennium Bug, allowed us to understand how such devices would become inseparable from the body and from the self in the synthesis of bios (life) and the hypermedia* of which the world wide web is the most classic example. Complimentarily, many other technologies, including biogenetics (which acts on the very code of life), nanotechnologies and robotics contribute to the flattening of the old nature-nurture dichotomy into a continuum from which the subjectivities that posthuman theories try to identify and define emerge. Since the whole of these tools act upon life, their spread and generalized use lead us to believe that capitalism is evolving toward a phase where cognitive becomes biocognitive.* However, we must not underestimate the fact that the metamorphoses induced by today's

9 A term created by Microsoft to indicate the early versions of today's smartphones.

technology produce a framework of the extreme commercialization that has for some time united economic power and political power.

It has already been estimated that "there is no longer an external that can limit the place of sovereign[10] "and today we might add that technological mediation intervenes in eliminating any delimitation: life and death, matter and logos, terra and cosmos, nothing escapes and this dynamic in turn causes new transformations. The interaction between these continuous mediations, which involve and form both collective intelligence as much as the individualist action of *homo-economicus*, arbitrates new modes of organization. We will now attempt to examine the aspects that most interest us.

10 Hardt & Negri, 2002.

NOMADIC SUBJECTIVITY FROM THE ERA OF BELONGING TO MIXED IDENTITIES

The rise and decline of belonging

We live in a new era: this is a feeling perceived more intensely by "pre-digital natives" than more recent generations. While the effects of integrating biohypermedia into the body are evident in cognitive productions and social relations in general and in P2P relations in particular, in other respects the vestiges of past modes of political organization remain, even if they are in decline. In Europe, the gap between the multitude and institutional forms of representative democracy grows deeper every day without this clearly indicating new forms of organization able to generate a constituent rupture.

Before moving deeper into the exploration of certain aspects of the current situation, we'll look for a few reference points in precedent modes of organization that may prove useful. The year 1848 is historically indicated as the beginning of this era. Defined as the "revolutionary spring," this was a key year of great uprisings against the restoration in Europe and the publication of the *Communist Manifesto*

("A spectre is haunting Europe – the spectre of communism…"). This explosion was fueled by social and political factors like the profound changes induced by the first industrial revolution that issued from a new repartition of labor, organized around technological innovations and new energy sources. With the mechanization of production, communication and information too were industrialized. In particular, the spread of printing and newspapers intensified, making a qualitative leap. This change in the media paradigm was essential for disseminating new ideas about social justice.

The turmoil, revolt and struggles could only correspond to the causes tied to the new way of perceiving and living the transformations imposed upon society and politics by industrial capitalism. The enthusiastic adhesion to ideals, typical of that era, was rendered concrete through the upsetting of previous social categories and gave birth to a new vision of society divided into classes. Classes not to be considered as ideological abstractions, but ones that were incarnated in uprisings, social struggles and modes of organization.

The workers' movement was born. It was led by the proletariat, because it was the workers who suffered the most traumatic vital dislocation. For the first time, labor was cadenced by machines and a disproportionate amount of life was spent in the factory. Both of these elements were destined to discipline the worker bios through a reflexive inurement, a necessary condition for rendering it productive and predictable. The reaction to this real subsumption to mechanical technology was, above all, sentimental and relational. The feeling of belonging to a new community where the strength to oppose this condition was the driving force behind social struggles that created the organization and that found its conceptual expression in Marxist philosophy.

In an attempt to understand the profound changes underway, a few thinkers developed ideas from the transition between the first, steam-engine industrial revolution, and the second, pushed by electricity. Nicholas Carr compares this passage to the current one that today leads to cognitive capitalism,[1] but it is difficult to equate entities that don't possess the same unit of measurement. In the two industrial technologies, the same entities are at play: energy, labor and power. It is energy that is transformed into power through the labor

1 Carr, 2008.

performed by machines, the essential element in a factory, even if the progress represented by electricity in respect to steam consists in rendering manageable, flexible and practical the production of labor power.[2] According to Simondon:

> Industry appears when the source of information and that of energy separate; Man [sic] remains the only source of information and asks Nature to furnish energy. The machine is distinguished from a simple tool because it is a *relais*: it has two distinct entrances, one for energy and one for information.[3]

On the other hand, Simondon laments the separation between manual and intellectual labor, weighing up the "industrial fatigue" of the iterative and fragmented regime of the unskilled worker – one he defines as "labor reduced to crumbs" – and the unhappiness of those who must conceive and invent without building or working.[4] This is a pertinent observation on a psychological level, but one that does not take into account the fact that these two kinds of suffering are situated in two different worlds: the worker's manual labor is certainly not paid as much as that of an engineer or inventor.

In the industrial revolution, the driving force generated by steam or electricity is destined, in one way or another, to substitute that of animals or humans in the manufacturing of goods: a decisive step towards realizing the expropriation of the diffused knowledge of pre-capitalist artisans, farmers, etc. A knowledge confiscated by machines in a mechanic and rigid way, thanks to the availability of this new driving force. What interests us here is, above all, highlighting the ties that are created between those who are subjected to producing through industrial machines and the emergence of a new need to belong, in particular the ability to recognize oneself in a collectivity that can defend itself from limitless exploitation.

The space of the factory is an essential element that is plotted over two axes: machines, to which men and women are subservient, and

---

2 In Italian, the word "lavoro" can also be defined as physical entity, similar to the English "work of art."

3 Simondon, *Mentalité technique*, 2006, 346 [our translation].

4 Simondon, 2006, 350.

the centralized availability of the driving force. Often, workers came to industrial labor after having been uprooted from country life. Being forced to spend most of the day together in the same physical location and the rest of it in specific neighborhoods adjacent to the factory is an important element in the origins of class identity. The feeling of belonging and adhesion to the workers' movement, as a bulwark of defense against submission is something that goes beyond participation in a community and becomes a shared project of a social life in common. A commons* made of struggles against exploitation and for the reappropriation of the means of production and the ideals of social justice and equality are the cement that permit the construction of new and powerful organizational forms like, for example, the political party and the labor union that, today, have become empty simulacra.

This need for belonging survived the 20th century and two world wars up until arriving in our era with characteristics that partially started to change in the '70s. As we mentioned in the previous section, during the "Glorious Thirty" after WWII, when the system was forced to concede a part of surplus value to the power of social movements, this sense of belonging was no longer limited to a single working class, closed up in the factory. In the meantime, with rising levels of education obtained thanks to social struggles, young proletariats accessed higher studies and became an essential part of the movement. The alliance of these two generations was fragile and ephemeral. The older generations were rapidly hegemonized by a worker aristocracy that, with salary increases once gained, entered into a consumerist spiral. A consumerism that was a powerful incentive to the ideology of property and an individualism that latched onto, and corrupted, the sense of belonging. Newer generations of workers and students possessed much higher levels of knowledge and biotechnological mediation already induced fundamental ruptures that swept them up and transformed them: birth control, easier travel, new media prospects like free radio, etc., were factors that came into play, creating an illusionary prospective of change that would also involve productive social relations beyond capitalism.

Even if it was less and less tied to the factory-space, a form of collective adhesion to ideals remained. A form that was still inspired by a social-humanist vision. Towards the end of the '70s, the hinge between the two eras, the great industrial entrepreneurship responded to this attack in various ways, including a change in the nature of large factories, introducing robots and automatization. In the '80s, Fiat's robotic factories were considered the avant-garde and this was the political and technological answer to the widespread social conflict in Italy of that time. Other aspects accompanied robotization; in particular, the new industrial organization of large factories presented in the first section where the assembly of subsets was delocalized, delegated and subcontracted to small and middle-sized companies organized across the territory.

This organizational turn corresponds to another resistance tactic of the workforce of large factories. Early on, already in '79, Antonio Negri speaks of a passage from the mass worker to the social worker[5] which would become the central figure in a mode of production that moved from the factory to the metropolis. The dispersion of workers and metropolitan production contribute to the undoing of the sense of belonging that the units of time, place and action in the factory had allowed. But the society of belonging is furthermore not exclusive to leftist worker movements; an equivalent conservative current existed, with a nationalism that overwhelmed part of the left, giving rise to the great, useless massacre that was WWI and, later, to mass adhesion to Fascism and Nazism.

Neonomadism: mixed identity society

In the brief space of two generations, we observed a profound anthropological change: when the anchors of belonging cede, an open sea where anything is possible appears. It is difficult to explain rationally how the collapse of a society founded on belonging left a territory lacking in berths where one can only be in transit or left adrift.

The progressive transformation of the nation-state into a global-liberal governance goes hand in hand with the compulsive construction of *homo economicus* in the sense of human capital given to it by Foucault.[6] The primacy of economic rationality forces us to vagabond in the particular nomadism of a hybrid space that breaks all moorings, not only ideological ones. Here we'd here like to analyze that space. Right along with the development of technologies, nomadism becomes an omnipresent reference point in media discourse.

5 Negri, 2007.

6 Foucault, 2005a.

Consultants and managers, young, precarious digital natives, immigrants and even 50+ housewives recycled through qualitative panels of detergents are transformed into contemporary nomadic figures according to the dominant stereotypes of marketing.

In spite of the increase in physical flows, a consequence of the industrialization of tourism and migrants forced to move by globalization, today's nomadism is far from Ellis Island[7] epics or the romantic journeys of *On the Road* or *Easy Rider*. Even if nomadism has become a condition that does not necessarily imply the movement of bodies, on a physical level many contradictions emerge, including those highlighted by Saskia Sassen,[8] according to whom it is precisely the forced migrant who aspires to settle and obtain a visa that allows them to finally feel stable.

Before venturing into the intrigue of today's nomadism, it is probably worth a second to take a step back and briefly touch upon nomadism at the dawn of humankind. For 99.99% of our history, which roughly corresponds to around two million years, *homo erectus* first and *sapiens* later, lived as nomadic hunter-gatherers.[9] Only ten thousand years ago, with the Neolithic and agricultural revolutions, did lifestyles begin to change, giving birth to sedentariness. Our prehistoric ancestors practiced a cyclical nomadism, naturally aimed at survival, that was in fact in harmony with the seasons and the natural cycles of animals and vegetables. Later, they became pastors and movement was tied to satisfying the needs of their herds.

Today, we can probably speak of a multiplicity of nomadisms that intersect, mix and that have physical, immaterial or virtual traits. In

---

7 Ellis Island, in Upper New York Bay, was the gateway for over twelve million immigrants to the United States as the nation's busiest immigrant inspection station for over sixty years from 1892 until 1954. The island was greatly expanded with land reclamation between 1892 and 1934. Before that, the much smaller original island was the site of Fort Gibson and later a naval magazine. The island was made part of the Statue of Liberty National Monument in 1965, and has hosted a museum of immigration since 1990.

8 Saskia Sassen, "La métropole du biopouvoir et la métropole de la biopolitique, gouvernance métropolitaine," speech at the seminar *Moltitude et Metropole*, Paris, 6 March 2006. See: http://seminaire.samizdat.net/spip.php?article181 [our translation].

9 It would be rash to predict such a duration for the nomadic lifestyle of the digital age.

a certain sense, contemporary nomadisms undergo a deterritorialization inherent to the form of cognitive capitalism:

> Capitalism is inseparable from the movement of deterritorialization, but this movement is exorcised through factitious and artificial reterritorializations. Capitalism is constructed on the ruins of the territorial and the despotic, the mythic and the tragic representations, but it re-establishes them in its own service and in another form, as images of capital.[10]

The process of artificial reterritorialization immerses today's hunter-gatherers into a flow that strikes the body-mind independently from its state and its dynamics without any continuous solution.

Biohypermediatic technologies intervene as a mediation tool for singularities with a space that is so pervasive it imposes an uninterrupted connection to multiple channels of interaction with a prevalent sensory orientation. These technologies participate in the reterritorialization and are functional to the modes of sociality of a nomadism that is individually more pronounced than in our original nomadism. The devices we carry with us are the current tools for gathering, and our weapons for hunting. Double-edged weapons because in such an environment, when we hunt, we can be hunted, when we gather, we can be exploited or looted. Our nomadic ancestors lived in a collective precarity because of the elements and natural circumstances, while the contemporary nomad has an individually precarious life due to the socio-political reasons that we will here attempt to highlight.

Contrary to the primitive societies of hunter-gatherers, today's nomadism doesn't seem to be guided by any logic; according to bioeconomic theorists it is entropic, in the sense that it inexorably and irreversibly increases the entropy of the system-Earth by subtracting energy from future generations and pouring disorder into the environment. In such disorder, there are no places set aside for resting or pausing: in the flow fueled by biohypermediatic channels, our senses are often saturated and emotions dominate, rendering any break difficult. Reality can be "augmented" or become completely virtual* but

---

10 *Anti-Oedipus: an introduction to schizoanalysis*, Deleuze & Guattari, 1972, 364.

the emergence of feelings diminish, thought remains superficial and doesn't crystallize, there is neither the space nor the time to get distance from the immediate, thus reflection escapes the ordinary.

The dynamic of crossing physical or virtual space has always contributed to a particular state in which one is immersed in an enveloping fluidity that hinders resting, reflecting and concentrating, especially when there is a motor that compulsively urges us on. When visiting the crown jewels in London Tower or the Cappella Sistina in Rome, there are human agents who urge us with words and gestures not to stop, not to photograph, not to speak loudly and so on, thus creating a continual flow that is taking us elsewhere but without any specific objective, without any determined stops or goals and that sometimes prevents us from perceiving the beauty that surrounds us. This is therefore different to ancient nomadism for which movement was motivated by the necessities of life: searching for opportunities that the territory presents in particular sites or seasons.

In the material world, the classic places of passage like stations, ports and airports are places of exposition, uncertainty and fragility. Attention is captured by the necessity to remedy an assumed lack of security. It becomes difficult to concentrate on any other activity, except maybe in the luxurious lounges conceived precisely for allowing managers and frequent flyers a productive continuity and for giving the elite a place to enjoy their exclusive privileges. The absence of pit-stops and breaks favors the absence of ties and, precisely so, any sense of belonging. This thrust is animated by automated algorithms designed for these functions and, although they are not human, they accelerate rhythm and guarantee the perpetuity of motion just like the agents at tourist sites mentioned above.

In movement, our senses are continually solicited, provoking emotions, i.e. the "complex programs that are largely automated actions formed during evolution"[11] and that our bodies are obliged to take. Today, emotions are voluntarily exalted everywhere, both in network spaces as well as in physical space, and it is indeed in emotions that the two spheres integrate and become complimentary. It is within this whole that the biohypermediatic sphere is born, where cyberspace, cybertime, organisms deceived by virtual reality are (con)founded with

11 Damasio, 2012, 136-137.

territory and real time between "hallucinations experienced consensually, and consensus hallucinated."[12]

The emotional machine set to work

"Live your emotions" and "Feel the passion" are imperatives abused in countless advertising slogans, from vacation packages and provocative perfumes to luxury German sedans with integrated software systems. According to the precepts of neuro-marketing, it is important that such emotions are as intense as they are "rare" and therefore dear to us and, as the preferred term promises, "exclusive." The luxury industry, most prosperous precisely in times of crisis, sells its products and services as a world apart of exclusive emotions. This ability lies in making the globalized multitudes believe they have the possibility of accessing it, and it works, as seen by the development of the counterfeit industry.

In marketing jargon, the emotions aroused by products (or services) are transformed into a "customer experience" (CX), a neuro-persuasion mantra. Here is a brief presentation of a discipline recently baptized "engagement marketing." The concept of CX was created by Morris Hoolbrook, a professor at Columbia University who, at the beginning of the '80s published an article with E. Hirschman entitled "Hedonic Consumption." In this work, the authors oppose the previously predominant vision of consumers as guided by rationality and therefore the consumer's experience is defined as "a subjective state of consciousness, accompanied by a range of symbolic meanings, hedonist responses and aesthetic criteria."[13]

Consequently, the client-experience of the product/service-emotion is extracted and analyzed in real time in the supplier's network in order to optimize investment returns; the system tries to transform life into a collection of experiences that are as profitable as the senses are saturated – so saturated that clients can no longer perceive the quality of the emotions and elaborate them into feelings. The commercializing, video gaming and Disneyfing of the real push us towards the state of simple organisms that are able to behave without mental processes; emotions but not feelings.

12 Gibson, 1986.

13 Holbrook, 1982, 132-140.

It has already been scientifically ascertained that behavioral rationality doesn't take place in people who don't fully feel emotions and sentiments[14] and thus disturbing questions arise regarding a society where emotions are so continuously influenced, manipulated and aroused by the obsessive prospective of financial rationality. What are the affects that feed and maintain the balance of today's economic rationality? Could they be the deep fear followed by the explosion of the economic crises that seem to be part of the very nature of capitalism? Or the reckless euphoria that tulip bulbs (the first speculative bubble in 17th century Holland) arouse before these crises?

The archaic god and the technological Leviathan – sacred techne

What has been discussed regarding neo-nomadism and transient modes of being has a counterweight (or contradiction) in new forms of absolute belonging that are manifested through archaic religious fundamentalisms. These extremisms are, in their various facets, more and more present – even hegemonic – in vast areas of the South. A conspicuous part of the global population is walking the road back to a sense of belonging that is even more archaic and binding than those previously discussed.

From a superficial point of view, fundamentalist movements seem to have in some way substituted those of Soviet-inspired national liberation from the Cold War era, a vision that doesn't however take into account the influence of the profound transformation that came with technological mediation. On the other hand, how can we explain the inconsistency of a North poised between the fascination and the threat of technological temptation and the archaic fundamentalism that, from the South, manifests itself even in western metropolitan suburbs? Simondon provides an interesting key for interpreting these profound contradictions.

In one of his main works dedicated to the modality of existence of technical objects, Simondon maintains, similar to what was written in the introduction, that the genesis of technical reality is part of human beings' relation to the world.[15] In addition, he adds that technicality is,

14 Damasio, 1995.

15 Simondon, 1958.

along with religion, one of the two simultaneous phases[16] that emerge in order to solve the problems presented in the *magical*, primitive original stage of our relation to the world. "Primitive unity," writes Simondon, "appears as a reticulation of the universe in privileged key points where exchanges between the living and the environment take place." These are places or magical moments[17] that are distinguished as figures distinct from the background of the universe. At a certain moment in evolution, we pass from the magic unity of these reticulations to the development of technical and religious thought that is "the organization of two symmetrical and opposite mediations."

In this doubling, or rather phase shift, key points in the world separate from the background to become a technicality that is crystallized in efficient and instrumental objects that function everywhere and at any given moment, while the background becomes abstract and is subjectified, personified in divine, sacred forms of religion. What prevents us from grafting the contemporary condition of a technology-religion dualism onto Simondon's vision? Simondon states that in the becoming of technical objects, key points of the magical, prehistoric world lose "their capacity for creating network and their power to influence reality that surrounds them from a distance." In this way, he refers to the technological mediation as we knew it until very recently.

Today, however, the situation has changed so drastically that we have put forth the hypothesis of this volume based on the paradigmatic leap in said mediation. A leap characterized, to use Simondon's terms, by the emergence of a context where today's technical objects (for example ICT devices and networks) are integrated with the "background" (the space-time of the universe), restoring, in some way, original unity. Such reconstitution obviously doesn't take us back to a world populated by magical places and doesn't entail transcendence but, contrary to what Simondon asserts, it can no longer be claimed that the technical object is "distinguished" from natural being

16 The phase must be understood, according to Simondon, not from a temporal point of view but from the point of view of the relation of phases to the physical, in which it must be conceived of as a relation to another or others and the whole of the phase constitutes a complete system (in our case, reality).

17 Many institutionalized and temporal vestiges of these figures remain today: holidays, vacations, justified with the excuse of the rest, "often compensate with a magical charge lost in contemporary urbanization."

in the sense that it is not part of the world. Quite the opposite, our hypothesis is that in human's "becoming machine," the technical object becomes a part of the living and this calls into question the vision of two mediations: the technical and the religious, counterposed as an indissoluble couple.

The basic framework from which technicality and religion were born at the dawn of human history is made brittle by a multiplicity of technologies that invade not only the political dimension of life, bios, but also the biological one: the vital breath of zoé. Evoking an extreme biopolitical case that acts upon the separation between bios and zoé and reduces life to "nude life," we can refer to Nazi thanatopolitics. Agamben reminds us of the Euthanasia-Program enacted by Hitler to eliminate incurable mental patients:

> [T]he program, in the guise of a solution to a humanitarian problem, was an exercise of the sovereign power to decide on bare life in the horizon of the new biopolitical vocation of the National Socialist state. The concept of "life unworthy of being lived" is clearly not an ethical one, which would involve the expectations and legitimate desires of the individual. It is, rather, a political concept [...] on which sovereign power is founded[18].

70,000 people were eliminated, of which 5,000 were children, in the span of fifteen months. The program was later abandoned due to the growing protest of the Bishops. The two doctors responsible for the program, condemned to die at Nuremburg, "declared they didn't feel guilty because the question of euthanasia would come up again." With the Aktion T4 program, the Nazis also widened their deadly action to all "lives unworthy of being lived."[19]

---

18 Agamben, 1995, 90.

19 Marco Paolini wrote and produced "Ausmerzen. Vite indegne di essere vissute." [Ausmerzen. Lives unworthy of being lived], a play that deals with Nazi eugenic theories and Aktion T4. This play was performed at Milan's ex-psychiatric hospital "Paolo Pini" in 2011.

"This is the story of mass extermination known as Aktion T4. T4 stands for *Tiergartenstraße 4*, an address in Berlin.

Today, for the first time, technology allows us to operate within the complexity that binds and separates bios and zoé and that, until recently, was indecipherable. In fact, like all mysteries, what unites life and death was the exclusive prerogative of religion and, in rendering it profane, we overstep the boundaries of the confines of religious thought and technical thought moves into the domain of the sacred. Paraphrasing Agamben, we could say we are facing a *sacred techne* that "is set outside of human jurisdiction without trespassing the divine." Therefore, from an archaic point of view, the civilization of profaning technology can be killed with impunity, as *homo sacer*, but not sacrificed.

On the other hand, this capacity to act upon bios and zoé opens many prospects including, in a positive sense, that of an era of hybridization that is not exclusively anthropocentric[20] that could give life to a non-capitalist, non-archaic ethics. Positive outcomes are not, however, obvious or to be taken for granted because, in this framework, technology is also the tool of the contemporary necropolitics practiced by biopower that, concentrated almost exclusively on the daily exploitation of life itself, creates inhumane forms of destruction. Inhumane are the new forms of a remote-controlled algorithmic death because it is delegated to automatons and robots like, for example, the CIA's drones that, in under eight years, killed thousands of people in Pakistan alone, including hundreds of women and children,[21] or the automatic sensorial strafing systems able to automatically activate themselves and shoot "intruders."[22]

These new forms of asymmetrical warfare, of which remotely guided drones are only the tip of the iceberg, are subverting the praxis, theory and ethics if not the very concept of war itself, as explained

During Aktion T4, around 300,000 people, classified as 'lives unworthy of being lived', were killed." Paolini, 2012, 5 [our translation].

20 Hybridization here isn't intended to support any particular current of posthumanist or transhumanist thought.

21 Already in 2012, there were more than 2400 dead according to London's "Bureau for Investigative Journalism": "March of the robots," *The Economist*, 2/06/2012 http:// www.economist.com/node/21556103.

22 For example, the automatic sensorial strafing systems like Rafael's Samson Remote Weapon Station, installed in Israel along the border with the Gaza Strip.

in the well-argued piece *A theory of the drone.*[23] More generally, the ecological devastation of the Earth is literally inhumane in the sense that it takes out a dangerous mortgage on the possibility of human participation in the future. However, now we'd like to focus our attention on the macropolitical consequences of questioning a reality founded on a technological-religious bipolarity. If technical objects, born from the objectification of magical places that emerge from the background of the primitive world tend to reorganize themselves in networks, pushed by cognitive capitalism and reconstitute a new unity, what are the consequences for religious thought?

The impulse of reticular technologies that reconstitute unity with the universe in the perspective of control and the commercialization of life and death calls into question the religious phase, breaking the previous balance. This condition influences all religions and, in particular, the three main monotheistic belief systems. Our hypothesis is thus that, subjectivizing and rendering "profane" the role traditionally allocated to the divine, technical capitalist thought unconsciously pushes the latter to regress towards archaic values by any means necessary. It is as if religious subjectivation tries to recover its primitive vocation of total need that it feels slipping through its fingers. In this regard, it is enough to recall the anathema of Pope Ratzinger – a theologian little inclined to the populism in vogue – against the "dictatorship of relativism." In looking for universal and absolute values, fundamentalist theologians are convinced they will find the original strength to contrast the invasion of technical, profane thought by going back to archaic values and ethics. This obviously doesn't mean that, for example, in Islamic theocracies the use of contemporary technology is denied but that, maybe unconsciously, they react against the supposed danger of a society that no longer has divinities to refer back to for ethics. This is common both in fundamentalist instincts as well as in the three monotheistic religions.

Thereafter, the force and effects of this phenomenon are different: in the areas of Christianity and Judaism, cradles of the new technological paradigm and where the decline of belonging strikes ideologies and religion, fundamentalism sometimes manifests with virulence,[24] though

23 Chamayou, 2013. For a realistic representation of drone piloting stations in the US, see *Good Kill* (Niccol, 2015).

24 For example, the somewhat ample social movement against the so-called "*Mariage pour tous*" [Marriage for all] (which extended matrimony to homosexual couples)

without assuming a driving or central function. In the great swath of the postcolonial south, from Morocco to Indonesia and where one of the great monotheistic religions, Islam, prevails, the situation is quite different. It doesn't seem surprising that facing western techno-biopolitical expression, archaic religious calls gain strength and increasingly radicalize. If post-capitalist social movements had managed to rapidly trigger new political processes during the Arab Spring, today we probably wouldn't be witnessing the wars that tear apart, disperse and take entire populations hostage, "collateral damage" of two asymmetric necropolitical blocs that fight in a downward spiral: biotechnological capitalism on the one hand and absolutist obscurantism on the other.

One of the expressions of the explosion of this antagonist equilibrium between the technical thought of cognitive capitalism and fundamentalist religion found its origins in the Middle Eastern wars to then spread globally. The two significant and rival arms are, on one side, suicide bombers and, on the other, Hellfire missiles launched from a remotely controlled drone that annihilate any form of life within a twenty-meter range.[25] The kamikaze and the technological angel of death are the incarnation of two deviations that attempt to destroy one another and us without any hope for victory.

If fundamentalist thought wasn't the archaic equivalent of Western neo-colonialist biopower which it opposes and if it had a minimal awareness of the impulses that animate it, it would have promoted Nineveh and Palmira as symbols of resistance rather than destroying them with several tons of TNT. In conclusion, nothing good will come of this war that opposes a simulacrum of god to the technological Leviathan origins of supreme algorithms attempting to subject the entire planet. Only a third path of constructing a common based on post-capitalist ethics can effectively counter this trend. The rest is a question of time.

Neurocapitalism: what Foucault could not predict[26]

We have already highlighted many aspects where the use of technology is the backbone of the metamorphosis induced by the fusion of

in France in 2014.

25 Chamayou, 2015, 120.

26 This passage partially re-elaborates the article "Biorank vs Commoncoin," Griziotti & Vercellone, 2014.

life and labor. Now we will analyze the shift in the dualities of nature/nurture and innate/acquired,[27] reexamining a paper presented by Michel Foucault in a lecture at the *Collège de France* on March 14th, 1979 entitled "The birth of biopolitics" and dedicated to American neoliberalism and its context. Foucault's intuition on the importance of the innate and the acquired in the valorization of human capital remains, but the philosopher could not have imagined the profound consequences introduced by technological expansion.

For more than a century, the speculative debate surrounding dualities infiltrated the domains of philosophy, psychology, medical research and human sciences and founded the disciplinary organization of society, including the Nazi regime, which made it the hinge of its destructive ontology. The society of control is instead founded directly on the manipulation of these two elements, innate/acquired and nature/nurture. For neoliberal dogma, they assume an importance in as much as they are fundamental components of human capital that, like every other capital, must generate income, the condition for earning time to live. In order to impose this economic rationality, the financial class that holds global power is therefore at the helm of the measurement processes of bios along with behavioral and genetic modification. Nature and culture, innate and acquired are struck by the technological tsunami where neurosciences, genetic engineering, nanotechnology, artificial intelligence and robotics all come into play. Our hypothesis is that in such conditions, these couples no longer constitute dichotomies: just as the separation between humans and machines is blurred, so does the separation between nature and culture, innate and acquired make way for a continuum.

In this new framework, we'd like to select Foucault's vision of neoliberalism as evidence and examine how technology structurally modifies these antagonistic couples in detail, demonstrating how they are evermore homogeneous. In the field of *culture* and *acquisition*, it is prevalently digital technologies and their biohypermediatic evolutions – moreover associated with the discovery of neurosciences – that

27 Nature versus nurture is the binomial that opened the modern debate on the relative importance of the inborn in respect to individual experience. Coined by Francis Galton during the Victorian age, this phrase was influenced by Darwin's Origin of Species. Francis Galton, *On men of science, their nature and their nurture*, Proceedings of the Royal Institution of Great Britain, (1874).

intervene in our modes of feeling, perceiving and understanding. They are used in increasingly subtle and articulated ways in strategies of influence, framing, business and governance.

At the heart of this new strategy lies Silicon Valley, where the algorithms and manipulation protocols of human neuroplasticity are conceived and implemented.[28] They are sugar-coated through stimulating technological innovations that facilitate the voluntary adoption of instruments of control in exchange for the illusion of individual liberty. Private initiative is also encouraged, accompanied and framed in the US by state agencies that leverage public funding; an example is the BRAIN program (acronym for Brain Research Though Advancing Innovative Neurotechnologies) to which the American government has substantiated significant investments. Under a therapeutic cover, this program perfectly lends itself to opening the door to re-elaborations and political modelling of neuroplasticity.

With such premises, more than a hypothetical Big Brother, if conditioned by the logic of economic speculation, the future seems to tend toward a meat-algorithm destined to classify the human and categorize it into compartments of integral exploitation, depriving it of its singularity and its capacity to produce an uncorrupted common. To better exemplify this tendency and to understand its risks, let's take a closer look at a few of the proprietary, secret and copyright-protected algorithms that play greatly influential roles. They generate knowledge according to a logic that uses predetermined criteria in order to establish what to show to whom.

The most well-known example is possibly PageRank, Google's algorithm that determines the rank and therefore the importance of a webpage and allows for a valorization of the visibility of a website on the net. This ability to create and dominate the market of classifying atoms of the network has made Google the most influential corporation of the digital age. Google has, for some time, concentrated on financial objectives being, among other things, the leading global advertising agency with an annual revenue of nearly 70 billion dollars in 2014,[29] smashing every pre-internet advertising record. The power of the oligopoly and the extraordinarily rapid accumulation of data now allow Google to pursue much more ambitious strategic objectives

28 See Terranova, 2013 and Munster, 2013.

29 Revenue relative only to advertising, not counting the group's other activities.

from an entrepreneurial perspective: to be the leader in the race to format humans in order to turn them into a pure function in the rise of neurocapitalism and in the transformation of life into commodity. This is not only a "soft" formatting, seeing that at the end of 2013 Google also bought Boston Dynamics; the company also specializes in designing military robots.

The PageRank algorithm is able to create surplus value from our network activity and reveals the potential of mathematical models able to approximate human behavior to the point of shaping it. EdgeRank follows the road opened by its predecessor. Facebook's algorithm intervenes more directly in relations by creating a classification based on which it automatically decides what should appear in the personalized News Feed of every member of the social network. EdgeRank establishes a relative weight of all our "friends'" posts using a certain number of parameters and criteria destined to quantify social relations. "Affinity" is measured by counting "likes," the frequency and type of contact with any specific "friend" who emits a post. Posts themselves are considered according to typology: for Facebook, the written form, which requires more attention and higher conscious cerebral activity, is obviously the medium that passes less, while photos and videos, which directly act upon senses and emotions without the mediation of cognitive elaboration, have a higher value. In addition, this algorithm takes into account time in a linear fashion: whatever is more recent is more valuable.

EdgeRank is only one important cog in a strategy that aims to create the equivalent of a privatized internet. Facebook's functioning is in fact based on the Open Graph protocol, "a technology that explicitly refers to the socio-mathematical techniques of graphs." According to Tiziana Terranova, "Open Graph represents a kind of privatized supplement to the Domain Name System (DNS) that defines the internet as a universal space of addressability capacity"[30]. In other words, it is a private extension that corresponds to Facebook's ambition to substitute the web in the daily use of its users and to make the existence of the internet as transparent as possible. The company's objective is to find a way that the universe of the network "members" is constituted purely through Facebook applications. Open Graph manages an immense, totalizing database that includes user profiles, photos, videos,

30 Terranova, 2015, 121 [our translation].

comments and, above all, users' all-important "likes" that Terranova considers to be Facebook's most important asset.

While Google's algorithm analyzes and exploits attention, knowledge and network behavior, Facebook's tries to measure P2P relations* in an aseptic fashion. The two algorithms have the common objective of creating a rank, an ambiguous term that has the double meaning of a position in a classification as well as a social class. In both cases, possibly more evident in the second, this rank is determined by examining and schematizing the facets of our subjectivity in order to measure and classify them.

While ranking algorithms operate on relativity (for example, the classification of an entity in a whole), rating algorithms, which create evaluations, work with absolute values. They are born from the financial field and today their use extends all the way to the point that financial rating agencies have more political power than any government except the US and, maybe, China. Parallel to ranking algorithms, ratings are largely used in the net. One of their precursors can be found in eBay, with its Detailed Seller Rating System that, in evaluating the activity of each small merchant and the occasional seller, became, in its own way, the Standard & Poors of the "poor."

As we saw earlier, the technological mediation of ICT has the capacity to capture both the body's biological parameters as well as changes in behavior and mood. In an opposite sense, they can saturate the capacity to assimilate and invest the mind with flows able to trick empathetic "mirror neurons"[31] and prepare us for the logic of human capital, where everything broken be analyzed and measured under the form of the investment that is destined to constitute it: from affect to oocytes, from genetic inheritance to professional skills.

This process shares a few similarities with what happened in the industrial era, i.e. when the factory implemented subdivisions between the living labor of the worker and the dead labor of the machine. Searching for revenue, neurocapitalism's algorithms go beyond the industrial machine, operating more than a subdivision, but an extensive corruption in the space-time of life. This mode of functioning sterilizes the richness of relations and directs desire, artificially saturating the senses and emotional responses. This is how the bulimic monster of Big Data is nourished. Fed with extrapolated data, it processes human

31 Iacoboni, 2008. See the Glossary.

relations, sensorial vibrations, feelings, gestures and behavior with algorithmic machines. This is just the preview of a path that, if left open, is destined to cross and change the thousand plateaus of human nature in an attempt to make it more and more homogeneous to the sole measurement of economic gain.

Capital and body

As we have seen, ICTs come into play on the cognitive aspects of subjectivity and other technologies currently under development act upon the "nature" and "innate" qualities, the first side of the "nature/nurture" and "innate/acquired" binomials mentioned before. These are techniques that act on the body, like in the case of genetic biotechnologies, and they contribute to breaking the rigid barriers in these dualities. In such a context, technological mediation is not limited to acting on processes that are reversible due to cerebral plasticity, but they are also at work in the domain of the irreversible domain of innate genetics.

Here we find ourselves in the field of living biotechnology where genetic engineers, using recombinant DNA molecules, can manipulate innate characteristics and therefore change the chromosomes received from the ascendant. Without wanting to negate or diminish the prospective and realities of therapeutic use, there is a clear risk that neoliberal rationality will tend to privilege the importance of genetic patrimony just as the aristocracy did with bloodlines. Finding genes that allow us to characterize the biological causes of complex disturbances and dependencies, from schizophrenia to addiction, may be used to justify a vision dominated by predetermination and predictability which would therefore be more functional to the logic of control. Meanwhile, the optimization of the value of human capital in order to increase it and, from an individual point of view, enrich the heritage of descendants by curing and preventing the risk of hereditary pathologies is a very real possibility. Foucault already spoke of genetic capital, but limited himself to the necessary investment that guaranteed an offspring "naturally" gifted with a good heritage like, for example, the choice of a genetically advantageous partner.[32] Foucault himself admitted that he was coming close (but not too

32 Foucault, 2005a, 187-188.

close) to touching upon a eugenicist argument. In the neoliberal machine, the technical tools for genetic manipulation, added to those for procreative control, can become the perfect complement to the qualitative ranking algorithms. They would finally close the circle of political, social and biological domination over the whole of society, extracting value from any activity, be it productive, reproductive or even unproductive.

Yet another important development in this field regards the other side of the coin of a new "industry" of living material. The research presented in the work *Clinical Labor: Tissue Donors and Research Subjects in the Global Bioeconomy33* bridges the gap of a deep reflection on the material mechanisms put to use by the pharmaceutical industry "that enroll *living* biology of human bodies in the processes of post-Fordist labor." In a review of this volume, Cristina Morini highlights the political use of technology in a new category of capitalist production, or rather:

> clinical labor, intending the process of extraction through which "the abstract and contingent imperatives of accumulation are put to work within the body." This clinical labor uses a manpower that is selected along race and class lines, in assisted reproduction technologies and in the sale of tissues such as oocytes and spermatozoids, generating a booming market of *bioproduction*. It must also be recognized that clinical labor (blood, sperm, embryo, organ and other "living tissue" donors) formally maintains a voluntary status, that of donating, itself based on bioethical principles that focus on freedom from coercion and on informed consent, but often ends up in practical terms being an efficient tool for facilitating "atavistic forms of employment contracts and discontinuous forms of reimbursement."[34]

"Clinical" labor therefore adds the "becoming raw material of the body" to the "becoming machine of humans." It would be superfluous

33 Cooper & Waldby, 2014.

34 Morini, 2015.

to insist on the fact that, in this process, the bodies involved in the "market" of human reproductivity are prevalently female bodies, "suppliers" that live in poor conditions and territories.

"In Europe, nearly all oocyte[35] providers are young women from Eastern countries who are trying to survive the insecurity of transitional post-soviet markets"[36]. Another case cited is that of Indian women who "rent" their bodies for gestational surrogacy. Here, we could also include research subjects and the enormous market of pharmaceutical experimentation where the problem of "professional guinea pigs" arises, especially in the large Chinese research centers. The government has implemented a policy aimed at "transforming China into an attractive destination for multinational clinical experiments."

Another context where living material and bodies become a source of power is the one tied to genetic code, just as the hypermediatized biologist-businessman Craig Venter has stated, whose symbolic deeds are told by Alessandro Delfanti.[37] Venter, after attempting to privatize the genome, converted to Silicon Valley's Capitalism 2.0 philosophy following a meeting with Google founders Larry Page and Sergey Brin. He abandoned the "proprietary" approach[38], at least regarding the human genome, to instead follow the now common path of those who sell services using open source information, which in the case of biology simply becomes open science.

Obviously there is an ethical debate in course, spurred by the commercialization of bodies and the temptation of eugenic applications of genetic engineering technology, but such debate in reality isn't able to limit the practices of the new forms of "clinical labor" that have therefore already become part of neoliberal bioproduction. We find ourselves facing the continual emergence of the risk of reducing the power and the richness of life to a *naked life*, a sort of zombie reflex that is malleable and permeable to everything and on which the violence of power is exercised. We are therefore at a crossroads: what is at stake isn't so much the development of evermore integrated science

---

35 An oocyte is a cell in an ovary which may undergo meiotic division to form an ovum.

36 Cooper & Waldby, 2014, 96.

37 Delfanti, 2013, 48-52.

38 This is an extension of the adjective usually used in the term "Proprietary software." See the Glossary.

and technology, but how to use them in order to organize social struggle and break free from the lethal model in which technology replaces ontology.

Skeptical generations

Following these observations on the biopolitical use of technology in the valorization of "human capital," we will now return to the changes induced by the passage from the society of appearances to the society crisscrossed by different generational behaviors. It is difficult to establish how much these profound changes in the way of being have been facilitated, or even triggered, in the process of vital subsumption where technology has today become a primordial tool. It cannot be denied that the "nomadic" state favors consumerism, overexciting the desiring machine to push it to impulsive/compulsive satisfaction or that make it less likely to sediment static or stable modes of organization. It is legitimate to ask if it doesn't contribute to the spread of increasingly frequent local and ephemeral unrest,[39] or if in such a context the immediate practice of revolt is more attractive than the attempts to constructively organize that are often co-opted by technocrats and bureaucrats of the infrastructure of governance.

Culturally, the root of the distant and "lone-wolf" attitude was already present in the punk movement. Although politically oriented towards the radical left, at least in Italy, a more introspective and disillusioned grain separated it from the previous era of belonging. In any case, we have seen how punk culture puts the individual body on the front line. After the involuntarily ironic decree of the "end of history,"[40] a Faustian pact was "proposed" to the generation of today's 40-year-olds: belong to the dominant model and to the resulting competitive behavior in exchange for advantageous professional placement. Instead the false myth promised wasn't honored: they underwent the "formatting" but received only precarity and the dismantling of the social state in return.

The following generation, "Generation Y," is more pragmatic and this isn't only due to its naturally higher level of technological

39 Regarding this debate, see the information and statistics published here: Antropologies du présent. http://berthoalain.com.

40 Fukuyama, 1992.

know-how. The bad example of their immediate predecessors, hoodwinked by lost enchantments and promises, together with the last fall of union rearguards[41], highlighted the impossibility of establishing a winning power relation with the same objectives and modalities of past struggles. Living in the cynicism generated by the obsession with economic rationality that hinders any real ethics, these generations are naturally skeptical about the human capacity to reconstruct a new collective solidarity while the dismantling of the institutions of previous generations is already underway.

The "shelter" provided by the net and video games where time passes without having to count it and, as represented in Zerocalcare's successful caricature[42] where surfing, (as opposed to immersion) is the metaphorical sport practiced, "distraction" is the most desirable activity in this unbearable world where time (money!) devours space. Virtual and evanescent sociality, fully immersive nights spent in a technical daze and chemical adjuvants extract digital natives from the growing aggressiveness of social Darwinist competiveness and from submission to measurement with the only prospective of a precarious future. The cooperative modes of production of the hacker community, on which much has been written and whose principle value lies in their affinity with these processes, sparked an undoubtedly interesting glimmer of hope, but it does not necessarily allow for the possibility of creating escape routes from the system's vital subjection.

The ontology of nomadic crossing therefore seems to be modeled around multiple factors combined because the transition is total: it is political, anthropological, sociological and ecological. As we have mentioned, it pervades every surface, seeps into every crack, uses cells, interacts with the processes and the map of the mind that are probably influenced and molded with dynamic forms at completely new speeds. In this context, capitalism retains the initiative despite the enormous difficulties intrinsic to the passage from the exploitation of labor to the exploitation of life and, in any case, is headed down the road of a rampant technological neuro-totalitarianism that (still) isn't explicitly claimed.

Currently, a debate is developing around the type of submission of labor that we are undergoing in respect to the classic Marxist

41 The last great union uprising in France was the movement against retirement reform in 2010.

42 See: http://www.zerocalcare.it/2013/01/21/la-fascia-oraria-delle-bermuda.

dichotomy of formal and real subsumption. Already in 2000, Carlo Formenti analyzed the characteristics of 21st century subsumption in a "Conricerca" interview:

> [...] we are facing a process that is apparently a return to a form of formal subsumption of capital: in reality, it is a process that is at the same time formal and real subsumption because the use of new technologies allows for the process of valorization to overlap real life, the territory, sociality, no longer separated like it was in the city-factory between the neighborhood of reproduction and the factory of production [...]. It is therefore a new, absolutely innovative form of capitalist valorization, destined to become increasingly capillary and efficient as the network is populated not only with individuals, subjects, collectives and traditions, but where "intelligent objects" themselves become a part of it. Little by little, as microchips are inserted, as intelligence is put into objects and they become potential network terminals, there will be the ability to innervate the territory and therefore recover information and data on behaviors, consumption, practices and tendencies: the minute practices of reproduction will be wired and, from this point of view, we all become *prosumers*, to use this ugly term. In the same moment that we consume, we produce, in the sense that we exchange information, we give the possibility to the other side to appropriate information and therefore constantly valorize the productive process. This is quite a radical leap...[43]

That leap has now been taken and contemporary subsumption is an unstable hybrid of that of the past. But this hybrid of formal and real subsumption from the industrial era, as Andrea Fumagalli affirms, gives birth to a "vital subsumption" that is more totalizing than the sum of its parts[44]. We could even define it as bio-subsump-

43 Forment, 2000, 2.

44 Fumagalli, *La vie mise au travail*, 2015, 59.

tion or bio-submission because it starts when life becomes labor and one could therefore speak of a transition from the theories of labor value and knowledge value to that of life value[45]. It leans towards absolute but, precisely for this fact, it is unstable. Despite the threat represented by neoliberal uses of technology, neuro-persuasion, genetic engineering and bodies becoming raw materials like in the dystopian science-fiction film *Soylent Green,46* this is rather like *Blade Runner* where we find an oscillating balance between the autonomous creativity of the common, on which cognitive business thrives, and the imperative of control through precarity and the privatization of intellectual property. This is what happens, for example, in the emerging sectors of the sharing economy, an invention of the common that capitalism sometimes (if not often) manages to corrupt and subject.

Bio-submission has the objective of orienting and transforming the ontology of transit into simple superficiality and acceptance of precarity, where the figure of the entrepreneur is the only possibility given. Nomadism is no longer the philosophy of life but a fictional pathway of continuous entertainment. In this spectacularized space, the dialectic confrontation transforms into an infantilizing Manichaeism much like Google's motto "Don't be evil." Google itself has worked for years with the NSA on Prism, a program for the global control of the multitudes, built with the blessing of the American government and later exposed by Edward Snowden, as masterfully shown in the documentary *Citizenfour.*[47]

45 Fumagalli & Morini, 2009.

46 Fleischer, 1973.

47 Poitras, 2014.

COOPERATION AND SHARING BETWEEN THE COMMON AND FINANCIALIZATION

Historical Nemesis of Capitalism

SUCCESSFULLY EMERGING FROM THE INDUSTRIAL ERA, THE LIBERAL system enters a new age orphaned of its archnemesis: an organized working class. The international division of labor, robotization and the metropolization of the factory destroyed the social relations of production and the antagonist equilibrium of the previous period.

Here, a global legal attack against "stable" wage labor begins; the last strongholds of the legal protections for workers fall, like the need for a valid reason to fire someone.[1] While it is clear that the mirage of restoring full employment in the biocognitive economy becomes slight-of-hand for waning unions, this does not imply that an endless decline is inevitable. But what new balance can be based on precarity?

Generalized precarity is the key passage from exploitation of labor to exploiting life as a whole. These new social relations are, however, unbalanced. How, then, in this situation, can the classic mechanisms

1 See in this regard the changes made to the respective legislation of the work in the so-called Job Acts in Italy and Loi Macron in France.

of consumerism and GDP growth be nourished? Here's how the economist Cristian Marazzi (2015) describes the paradoxical situation of a capital that suffers from its own victory:

> When capital has destroyed the working class as we have it unknown – above all through eliminating fixed salaries and contracts, as well as other capillary measures for job precariousness – the very possibility of not only creating liquidity, but integrating it into the economic circuit is lost. Money is created to monetize wages; when wages no longer come in the form of working class bargaining, the doors to an integration of liquidity that goes all over, that creates rent and income not as a lever for consumption but as unproductive, ultra-concentrated wealth in the upper classes – the famous 1% – that cannot trickle down towards the lower classes. There is therefore a serious problem created by the destruction of the institution of wages, which makes monetary policy fuel financialization and the upwards concentration of wealth. In this sense, the problem is structural.

Re-adapting the enclosures system,* i.e. the fencing of common lands situated in the new cyberspace or biohypermedia dimensions,[2] is part of this strategy, especially when the services offered are attractive (like in the case of Facebook) because they capture the capacity of new technologies to meet social needs. However, in the way they are proposed and implemented, they seem to constitute a further step towards a totalizing subjugation of life.

A house of cards...

Before widening our gaze and presenting our hypotheses about possible organizational approaches – necessary in the face of global threats – a handful of considerations on the contradictions between

2 For example the fences of Social Networks or those created by the unconditional reflex to use Google for any search. Then there are those of online commerce, first Amazon.

the resilience and the unsustainability of cognitive capitalism seem in order. One wonders how it is possible that the multitudes, with their potential intelligence capable of building complex and sophisticated constructions, find themselves harnessed in compelling and inextricable situations of precarity and unemployment, and why they suffer social cuts and the penalizing divarication between rent and salary.

At first glance, the capacity for control expressed by biopower, though reticular and pervasive, may seem like a house of cards, ready to fall at the first breath of a movement animated by the consciousness of its own capacity for autonomy. But appearances can be deceiving; we are forced to recognize the non-linearity of any possible escape trajectory. In the previous section, we formulated a few hypotheses on the ways in which social relationship issues in the knowledge economy are harnessed by the hegemony of cognitive capitalism*; we will now widen our gaze to the forms of cooperation of the common.

For example: will capitalism succeed in transforming the ever-increasing ecological destruction into new "markets" of "ecological" spaces reserved for a caste who can afford it? In the tradition of Orwell or Huxley, many works of science fiction and, later, the Cyberpunk movement, evoke traumatic evolutions or ruptures and the establishment of distressing and sometimes apocalyptic dystopias. It's no chance therefore that, in certain scenarios, a part of humanity is rigidly confined in hyper-polluted environments. In the film *In time* (Niccol, 2011), for example, we see a possible dystopia of the neoliberal integration of innate and acquired, i.e. a society where currency corresponds to life time. In fact, people are genetically modified not to grow old, but can only live for the duration of the capital-time they possess: a few privileged "rich" live a life without limits in the rare, unpolluted places, while others live in a degraded land, fighting for survival.

Already today, certain Asian or South American megalopolis are more than a prefiguration of this kind of hell. The air is unbearable and, according to the British journal *The Lancet*, in 2010, in China alone, 1.2 million people die prematurely due to air pollution. A self-replicating castle of cards? Schumpeter's concept of "creative destruction"[3] seems to be lost in an almost geometric acceleration vortex.

3 Schumpeter defines "creative destruction" as that in which, under the boost of innovation and the resulting selective process, many companies disappear, others

How many first generation ICT manufacturers are still around? Who, for example, remembers the first PC manufacturers[4] and publishers of database management software from which what we now call "big data" was born?

In the best marketing tradition, every card in the castle is marked by a logo. Take, for example, the case of eBay, the global online auction, now more than twenty years old. Undeniably, the idea of being able to sell used goods online is a positive development rendered possible by the net which, among other things, makes it possible to recycle, extend the use-value and reduce waste and e-waste. eBay has succeeded in acquiring global leadership in this activity (excluding China) and, by taking a high percentage profit from the sale and purchase of internet users, it enabled its former financial partner, Paypal, to become one of the most dominant payment methods on the net.

Although it is technologically possible to organize a production of the common that is not subjected to exploitation and a collective life not subject to rent extraction, companies seem able to control and dominate these activities. In this capture of common cooperation, finance plays an essential role. It hierarchizes, promotes and directs initiatives based on the business plans that are most attractive in economic terms through a differentiated reticular infrastructure that goes from the traditional banking circuit to crowdfunding. These realities mustn't make us fall into a Manichaean vision; we cannot forget the capitalist technique of adapting to real needs and propose new, attractive features. In an interesting inquiry, the young French journalist Jean-Baptiste Malet (Malet, 2013) revealed, after having been hired by Amazon, how this venture has become the symbol of regression in labor conditions in countries such as France, Italy and Germany at the beginning of the 21st century.

Despite this information, the complaints and the many strikes (above all in Germany), why do we continue to use this American website when we are well aware of the real working conditions imposed? If it's true that Amazon's prices, supply and quality are often tantalizing, why should we have to pay such a high social price for such "efficiency"?

---

are born, and others are strengthened.

4 IBM, for example, sold its PC division to the Chinese manufacturer in 2005 Lenovo that later comes to the top of the world rankings.

...of many surprises

When a wave of protest advances, driven by expanding poverty and precarity, then to repress it from the self-replicating castle, control comes out as robocop in flesh, bone, plastic, iron and silicon, trained to obey and exercise violence. Governance's motto is stopping any revolt at the bud and catch the slightest sign of weakness in social movements which, among other things, reveals a deep fear of losing control due to disattention. Movements in Italy and France and the initiatives against the useless public works projects that cause further ecological devastation have been severely repressed, both legally and militarily, with fatal levels of violence.[5] Or, like when the 2011 Occupy/Indignados movement was immediately suffocated after attempting to set up camp at *La Défense*[6] with the deployment of armed forces so disproportionate one might have thought the French oligarchy's terror to any collective expression had still not been placated since the Commune... A concern no doubt amplified by the fear of a "global" common no longer limited to one city thanks to networks. In Spain, on the other hand, despite the inevitable drop in tension, the Indignados movement has been able to take root more deeply throughout the territory and, a few years later, the semblance of the capacity for an electoral mobilization has led to victory in at least a few local elections, including the *Barcelona en comú* list (which confirms the centrality of the term "common" as we will discuss below).

The apparent transience of social movements and the shocks that have transfigured many Arab countries and the Middle East[7] must not mislead us. For those who continue to use the deforming glasses of the period of stable political identity, the situation appears desperate. However, if those old illusions are left in the past, surprising

---

5 Assassination by the French Gendarmerie in October 2014 of the militant ecologist Rémi Fraisse demonstrating against the Sivens dam designed for the intensive cultivation of corn for animal food, by destroying a large natural area. No penalty or suspension or serious inquiry was carried out and the non-commissioned officer responsible for the launch of the "offensive grenade" that killed Rémi Fraisse "resumed service" after a simple interrogation.

6 Paris' Business district.

7 Without prejudice to the case of Tunisia where, instead, the 'Arab Spring' has established a change that appears to be more stable, although now subjected to strong turbulence.

perspectives can be seen. From the ashes of an apparent resignation, a collective unconscious with a strong potential for autonomy still exists, although it has not yet found the capabilities of structuring new institutions of the common. The creation of affinity organizations can be slow and uncertain: nomadic subjectivities have perhaps less patience for prolonged attention and hierarchy, and are much more tempted to compromise due to the fact they are subjected to different neurological bombings that make sustained concentration uncertain.

However, the network makes it possible to connect and make possible viral contagion, but the uncertainty about the forms that this new autonomy can assume remains strong, and, excluding unforeseen accelerations, the time necessary for rooting these movements deep into society remains long-term. The consumerist hedonism that favors individualism has marked the last generations and continues to do so. Never as before, the term "common" and its variant "the commons" is topical, whether referring to goods or to the construction of a "common" as an alternative to the two poles of the public and private sectors, or the "new commonwealth" of cooperation. It is probable that this movement of the common is born also in response to the spreading logic of an individualism that is increasingly pushed to the extreme.

Despite the differences in meaning and political orientation of all these variants, the irruption of this term into the debate and public discourse is a sign of the times, dictated by necessity. Facing a financial system that represents the "most abstract and detached form of command"[8] at the very moment it concretely engages life as a whole, the common represents the need to find an indispensable form of cohabitation and mutual support that makes life worth living.

Affinity hanging in the balance

The temptation of looking towards the past with regret is a generational stereotype for those who spent the first part their lives in the age of political identity. In this case, we have likely to catalog nomadic ontology as a regression, understanding only its veneer of superficiality.

8 Toni Negri, "From the end of the national left to the subversive movements for the ropa," www.uninomade.org/dalla-fine-delle-sinistre-nazionali-ai-movimenti-subversive, 2013.

In a change of this scope, it would be misleading to limit our vision to only this aspect, and moreover, superficiality is often induced as a facilitator of an integrated and accepted precariousness.

Capitalism has now abandoned all ethical references, including the now obsolete Weberian Protestant morality. Neoliberalism is a managerial approach to economic organization where technology tends to be a transcendental value. On a collective level (but not on an individual one), we are lightyears from the dawn of liberalism that, in some way, came about as a response to a disciplinary logic that produced famine. Today, on the contrary, it has given rise to multiple wars and planetary destabilization, starting from the Middle East. In economic terms, it produces post-industrial famine and its German variant, Ordoliberalism,* more realist than the dollar king, that generates the same in Greece and in the other countries of southern Europe. Historically, capitalism made entrepreneurial spirit its flagship; today this sense of initiative remains strong in field of startups where it interacts both with the classic system of venture capital,* as well as with the more recent crowdfunding. The long-term battle of opposing this concept also passes through a repositioning of these roles, important in the development of forms of autonomous production of the common.

The hierarchy of governance no longer requires unconditional collective and explicit accessions, but rather applies a *laissez-faire* attitude on an individual level. Facilitating a superficial subjective approach, it is easier to gain affinity with the emerging modes of autonomous organization. Why fight it when externalities created by independent cooperation – like free software – can be used for profit? Why not cultivate an intimacy that allows the reigns of exploitation to be strategically positioned around these modes of production? The circuit seems closed, and one gets the impression of returning to a magical world where biopower occupies every key position, thanks to technology. The Foucauldian ambiguity of a liberalism as more a political technology than an ideology – successful because "it does not break away from the game of reality with itself"[9] – finds one of its most significant expressions in the management of bio/technological startups.

The creation of start-ups is strongly encouraged and is presented as an attractive and "autonomous" choice but, for the large majority of

9 Foucault, 2005b, 47.

new entrants on the labor market, this thesis is translated into practice and, when it does work, results only in precarious labor. Others become precarious to escape the psychopathological conditions of labor in the large sectors of cognitive labor. This situation leads directly to considering oneself human capital that generates rent and, if one does not belong to a class where an "investment" has been made, then one becomes an insignificant asset with little "value" and few "returns." If, instead, an investment were made, perhaps on credit, then this "asset" would belong to the debt pool, a further and omnipresent biopolitical tool for control.

Fossil wars and false prophets

Cognitive capitalism pursues a policy of scarcity, adapting it to a new technological context. In the essay on "The Zero Marginal Cost Society" Jeremy Rifkin (2014) forcefully denounces the actions that limit what, thanks to the internet, should be abundant.

In addition to condemning the artificial shortage caused by copyright, the theorists of a society of abundance without limits extend this principle to other sectors, such as logistics and energy. The themes developed in this essay evoke the return[10] of collaborative commons and, in particular, an energy common where the advent of an internet of renewable energy, produced by multitudes, is envisioned. These hypotheses are undoubtedly attractive and well documented but they also seem to underestimate the rigid central control exercised in the field of energy production and distribution in public and private capitalism that has constituted an oligopoly in this field up to now. Today, "Smart Grid" technologies,[11] or intelligent distribution networks, are used in such an opaque way and are insufficiently aimed at

---

10 According to Rifkin in the feudal period agricultural production was organized under form of Commons, until the advent of the enclosure, the enclosures of the lands introduced the modern concept of private property. See also "Enclosure" in the glossary.

11 In smart and electric engineering, telecommunications is the term used for a power distribution network driven by information technology in such a way as to allow us, in theory, to manage the electricity grid in an "intelligent "under various aspects or functionalities in particular for what concerns the efficiency of its distribution.

reducing the greenhouse effect and the elimination of nuclear energy. The development of renewable energy spread across in the territory and managed through the network would allow for both a low-cost supply of electricity as well as a substantial reduction in pollution. Unfortunately, this is not the policy of European governments or the United States. In France, for example, with its highly-centralized administration and where in 2015 more than 70% of electricity still originated from nuclear power, EDF (which virtually holds an energy monopoly) decreased the purchase price of solar energy for individuals who "manufacture" it on several occasions, making any profitability unlikely.[12]

Our doubts on the affirmation of an economic paradigm that makes the private sector "less practicable" in the field of energies are corroborated by other trends. Take, for example, *smartmeters*, the new electric meters that should provide information and give a higher degree of independence in reducing consumption. The logic adopted by large energy production and distribution companies seems rather inspired by the philosophy of control: in France, EDF's new "Linky" meters are designed to identify the number and exact type of devices connected, including the television channels. Obviously, users will not have access to their own data which will instead be made available to the supplier. Likely to add insult to injury, if legal challenges are not honored, these spy-meters will also be charged to users who will have to pay a monthly rental cost for a closed and "proprietary" system.

There is no doubt that the energy oligopoly favors fossil fuels (just look what happened with the race for shale gas in the USA and Canada), even though this oligopoly is also subject to geopolitical tensions. Pandora's box of fossil fuel wars was opened by Bush's clamorously defeated Texan oil lobby, but which left behind the desperate and dramatic situation in the Middle East, with millions of victims if we count the dead, the wounded and the refugees. With widening conflicts that have now reached the borders of Europe, and in the

12 See, in this regard, the investigation by the French Consumers Union in 2013. At the price paid by Electricité de France (EDF) self-produced energy, the profitability threshold for a photovoltaic system is between 10 and 48 years, in other words, the investment can be considered a loss. In 2014, EDF still lowered the purchase price of energy produced by private individuals. http://www.ufcquechoisir-paimpol-lannion.org/index.php/ electricite1/243-photovoltaique.

confusion of incoherent and inconclusive American diplomacy, the only intelligible fossil fuel strategy is that found in Gulf countries, who open extraction taps and drive down (perhaps beyond their own predictions) the price of oil, partially putting expensive American-Canadian extractions out of the market.

The energy oligarchy also tends to monopolize the production of renewable energy, favoring large solar, wind or other plants. Rifkin, along with the economist Piketty, the best-selling author of five hundred pages of economic science (Piketty, 2014), imagines the decline of the current system without any conflict. In the case of Rifkin, a series of free simplifications, shortcuts and conclusions as optimistic as they are suspicious are derived from his naive vision. The energy sector, and in particular extraction, is central in the history of industrial capitalism as we have seen, even to the extent of starting a lost imperial war in Iraq, of which we still suffer the consequences. Given these precedents and the strategic interests that come into play, it would be ambiguous to pretend that the hard core of capitalism, as is found in the control of fossil energies, magically turns into an idealistic energy commons without a fight.

The processes of economic financialization allow for a measurable generalization not only of goods but also of services and therefore already constitute the system's response to the fall in cost margins and profitability in industrial production. The prosumer, whom the false prophets of the "end of labor" would like to pass off as liberated producers, correspond rather to people who, more or less unconsciously, provide free labor for profit-driven companies. Taken for granted, the end of wage labor merely implies the affirmation of precarious labor that is unrecognized or unpaid, but that creates measurable value through finance.

There are also sectors in which the end of labor, due to technological advances, are not applicable in the short- or medium-term, like in sphere of "human-to-human production." In this sector, we find welfare services like education, healthcare and the right to culture, once acquired through past struggles and that, in Europe, have for years been the target of cuts and privatizations. The underlying political will is to gather these services into the private sector and make them as profitable as they are in the United States, where it is normal for those who have the money to be taken care of and those who do not to die. This story is told in many American films, like *John Q* (Cassavetes, 2002) and *Sicko* (Moore, 2007).

In the services that remain public, a certain Taylorization is carried out against those who work in healthcare in order to decrease the costs of "production" and the amount of employees, resulting in the deterioration of doctor-patient relationships. Robotization is also ongoing in this field, even if limited to specific niches. In Japan, for example, the country with the oldest population in the world, robots are progressively being used for the surveillance and entertainment of the elderly, disabled and other categories of assisted living.

The limits of hacker ethics

The hacker movement that, as previously discussed, finds its roots in the political and cultural context of the '60s and '70s, represents the main form of the autonomy of the collective intelligence opposed to the financialization of life and labor. In Pekka Himanen's essay *The Hacker Ethic and the Spirit of the Information Age* (2007), as well in many other writings (such as those of the anthropologist Gabriella Coleman), entertainment is celebrated as a fundamental condition for hacker labor. The etymology of this word is "*tenere intra*," evoking a state of delay, the passing of time, certainly in a pleasant way, but without any particular purpose. We are far from what André Gorz (2008) evokes when he describes liberated labor as "self-production" or "self-realization." In self-realization, we could include the desire to take up projects, organize cooperation, valorize skills or lead projects. What then distinguishes the capitalist entrepreneur from the hacker? Perhaps the contradiction between the stress derived from the competitive-profitability imposed by a hypercompetitive "market" and the impossibility, in this condition, to have fun while working.

Digital sector companies draw their lifeblood from startups and the hacker multitudes, an osmosis that has been functioning for decades and that is achieved by cultivating cultural affinities with the Californian stereotypes like the casual wear of the now legendary cocoon campus. Googleplex is in fact designed to satisfy consultants' and engineers' every need and whim, encouraging them to stay as long as possible in this precious cage to hatch golden eggs. Devices that facilitate this climate of fun, typical of the hacker environment, play a primary role in maintaining the affinity between the latter and Silicon Valley's digital capitalism. Google's policy of leaving every employee

20% of their working time for personal initiatives is certainly a good one, but the company alone captures the financial results.

Contradictions, however, begin to emerge: in San Francisco, the *Heart of the city* collective[13] organized a protest against the Google-buses that make use of public infrastructures. This protest concerns San Francisco's galloping gentrification. To make room for the company's well-paid consultants[14] and the neighboring Silicon Valley employees who receive dizzying salaries, rent and the expulsion of the less affluent and "less white" population are on the rise and San Francisco risks losing its alternative soul. This seems to have been confirmed by a recent local vote that rejected limitations on short-term rent after an undemocratic campaign affected from the financial power of Airbnb, a company that injected more than eight million dollars.

Regarding this issue, many questions remain: what is the relationship with money that emerges from hacker ethics? Which compromises are acceptable? What concrete differences exist between hacker cooperation and those of a software company? The logic to which the startup founders must submit is the consequence: success consists in being able to sell their activity to a corporation that wants to take possession of concepts, patents and new technologies, except for the very rare cases when startups become corporations themselves or when they fail (i.e. the vast majority). In the ICT field, there are important areas and organizations that resist: besides Richard Stallmann's historical and previously mentioned Free Software Foundation (FSF), we could also mention the Electronic Frontier Foundation (EFF) or the Chaos Computer Club (CCC) in Germany. In Europe, often such associations have more politicized and anti-capitalist positions than in California and the United States.[15] They hold periodic meetings, like the annual Italian Hackmeeting, started in 1998.

In the Bologna Hackmeeting of 2014, for example, following Snowden's NSA revelations, the theme was digital safety and techniques for protecting against electronic espionage, while in the seminar dedicated to cryptocurrencies, an explicit critique of the role of

13 See http://www.heart-of-the-city.org.

14 Google is the best company paying its employees in the US $128K / year on average.

15 On the genesis of these instances and associations see "L'âge du Faire" cap. 2 (Lallement, 2015, 75-105) and *Networked Disruption* (Bazzichelli, 2013).

bitcoin was developed and the possibility of creating a currency of common resistant to financial speculation was discussed. The use and exploitation of free software and open source in the for-profit economy induces the development of new specializations, increasingly in demand by large companies or large projects like, for example, Open Source Consulting. These specialists can precisely dose the percentages of free code to be used in such a way as to maximize return on investment. These consultants are used by the vast majority of the digital service companies to maximize sales of their services. "The shift from ownership to access" (Rifkin, 2014, 317), which will be discussed below, definitely contains innovative aspects, like the reduction of waste, but since it is run on a logic of profit, it does not affect financial control over society.

The referee of sharing economy

Capitalism in the biocognitive era seems to be agile in moving the cursor from the extraction of value to the product and services as well as positioning itself in new channels of networked social cooperation. Is there any way of liberating P2P* (peer2peer) cooperation from financial rationality? Is there a vanguard capable of exerting a hegemony similar to that of the working class in the previous phase? Or is is possible to reform contemporary capitalism and find acceptable compromises? At first glance, it seems that all hope of reforming ordoliberalist* European command was lost with the 2015 summer crisis and the yoke to which Greece is now subjected. Financial governance flaunts a panoply of incentives that branch out into the territory and are able to draw in cooperative initiatives, shifting the balance in its favor. However, when this is not enough, it beats its iron fist and imposes its will.

The ability to maintain such a rule in this new context where fixed capital is now largely concentrated in the intellect of workers, capitalism may seem at first sight much weaker than it was at the time when large industrial investments were necessary. However, we must not forget that there are inherent limitations to the concentration of fixed capital in the General Intellect. While individual means of production in the digital age are easily accessible, the reticular functioning of infrastructure implies handsome material investments. If stocks were abolished, as some economists suggest, the technological foolishness of High Frequency Trading would be unnecessary and, in the same way, if an alternative cryptocurrency of the common replaced the proof-of-work system, then it would avoid the waste induced by the current generation of speculative cryptocurrencies, like bitcoin.

It seems difficult to decide what the future of data farms* will be. Today, they are the raw material of the cloud* which contains big data. Although they are energy guzzlers, it seems unlikely we can do without them because, in fact, they are essential components of digital energy. The problem is that data farms are the empty containers that multinationals manage to privatize digital content.[16] Big data is not in good hands since, for example, Amazon – beyond being the world leader in e-commerce[17] (and champion of the neo-exploitation of workers) – is also in pole position for cloud services. The open data movement, considering most of big data as a common good, is opposed to this excessive privatization of data:

> Numerous scientists have pointed out the irony that right at the historical moment when we have the technologies to permit worldwide availability and distributed process of scientific data, broadening collaboration and accelerating the pace and depth of discovery [...], we are busy locking up that data and preventing the use of correspondingly advanced technologies on knowledge.[18]

Given their strategic importance as containers, data farms should then become part of the common, like the internet or the web. In an environment where data appropriately becomes common, *blockchain* technologies – which we'll examine later – and crowdfunding applications free from the control of finance are part of a potential future that could allow for the creation of trust in the institutions of the

---

16 For example, in the case of Facebook in the general conditions of the contained use nuti once published by the subscribers (such as photos, videos and comments) diventa- no ownership of the site that claims to another not responsible for the contents themselves.

17 $136 billion turnover in 2016.

18 Statement by John Wilbanks, former VP Science, Creative Commons, https://learn.canvas.net/courses/4/pages/open-data-in-more-detail

common, replacing the current subordination to the motto "In God We Trust."[19]

Learning the common, the common of learning

In many countries, Fordist institutions are crumbling, and public education is certainly among the worst of these. There can be no shortcuts when speaking about the threat which seriously weighs on our future generations. Statistics from around the world highlight how cognitive capitalism is transforming higher education in every country into a for-profit business into factories that produce indebted students.

Nor will we delve into the prospects, made possible by network technology, of rendering teaching and learning more accessible, not to mention the realm of self-learning – a trend that fully builds on autonomy and the self-management of knowledge and understanding. From Montreal to London, from Chile to China, we see student movements against the school-enterprise and a factory of temporary workers who are struggling to oppose long-term indebtedness. This is the reality of things and, without seeming too suspicious, it is unlikely that quality higher education can magically become free and affordable for millions of students just because the usual prophets say that, thanks to the internet, new corporations will diffuse knowledge *urbi et orbi*. When Silicon Valley capitalism is guiding the so-called *mooc* (massive open online courses), we should fear that the saying "if the product is free, you are the product" is becoming a reality.

We will therefore simply mention the increasingly frequent birth of experimental, autonomous and self-managed learning projects that do not aim to churn out ready-to-consume cognitive workers. We will limit ourselves to four modest, grassroots experiments that are, perhaps, a sign of a more widespread and growing underground. Regarding Sujonomo N in South Korea, as described by Franco "Bifo" Berardi, who visited in 2013:

> When asked of their activities, many of them define themselves as independent researchers and consider Sujonomo the place where their work can develop and be made public. The neoliberal privatization

19 Motto that appears on all Federal Reserve tender.

of the university has advanced quite far in Korea, as I verified during a conference convened by groups of students, in which they asked me to talk about the European movement against the Bologna Charter. Consequently, a growing number of students and teachers in this country are abandoning a university that is more and more expensive, more and more submissive to business purposes, and less and less useful for finding work; they thus create self-education and independent research spaces.

The Sujonomo's activities are varied: [students] meet in two spaces in a working class area of the city. In one of the two spaces, they cook and eat together, do yoga, and listen to music. In the other, they hold seminars and collective readings. During the time I was with them, beyond my seminar there were three others: one an ongoing study group on sexuality in Foucault, one on the thought of Benjamin and another on mathematical issues related to digital technology.

Sujonomo does not claim to act as a political entity, but it works quite usefully as an independent center dedicated to the self-organization of cognitive labor (2013).

Commonware

An Italian experiment, born within the Uninomade collective, Commonware has remained active even after the former was dissolved. It defines itself as follows:[20]

Commonware is an apparently cryptic and deliberately ironic name. We chose it to mock the instructional packages of university companies, so-called courseware, inverting the sense to indicate free, social cooperation. We chose it, above all, to name our challenge: building an "education of the common" that is at the height of the new composition of living labor, immersed in the

20 http://www.commonware.org/index.php/progetto

struggles and the materiality of collective behavior.

The Maîtres Ignorants[21]

In 2012, after the ebb of the Indignados/Occupy movement in Paris, a group of people who had participated in this movement launched an initiative of self-training on real democracy. It looks like this:

> It's a self-training center for democratic practice. The intent of our project is to ensure an inclusive and horizontal nature like an assembly, to know and to master the tools of discussion and decision-making on the internet, enabling collective intelligence to express itself while respecting individual autonomy and avoiding the pitfalls that lead to sterile opposition or even sterile unanimity, rather than to the development of common projects... all this requires not only the will but also the methodology, techniques, and skills we need to acquire.[22]

The experience lasted three years, teaching the direct democratic practices of organizing courses on common currency, the hacker movement, makers, etc.

In conclusion, we would like to present Unipop, an Italian experience developed as part of the Milan community center "Il Cantiere":

> The "People's University" is an experiment in producing and exchanging knowledge through courses, debates and meetings of self-education. It was born from the experience of the Italian language school for immigrants "Abdoul Abba Guibre." The understanding of a language is an essential tool for social interaction and access to rights, just as it is for deepening

21 Title from the essay: J. Rancière, *Le Maître ignorant: Cinq leçons sur l'émancipation intellectuelle*, Fayard 1987 where you can see the experience and ideas of Joseph Jacotot, teacher and French revolutionary of the late eighteenth century who questions the dogma of the intelligence on which inequality still relies on teaching.

22 http://www.lesmaitresignorants.org

historical and contemporary political debate. The understanding of languages and the conscious use of new technologies are the main tools for understanding the world, let alone indispensable for taking a stand. Self-education is a practice of resistance for organizing our lives, today's conflicts and for imagining another world for tomorrow.

The systematic destruction of schools and universities promotes the passive acceptance of models that preclude any possibility of imagining the future: the places of decision-making regarding our lives are increasingly distant from any form of participation and democracy ends up being a screen behind which a handful of rich and powerful pursue their private interests. Only collective intelligence and sharing knowledge can find the best ways to fight the plundering of resources and the abuses and devastation of our lives. UniPop therefore aims to be an institution for understanding what ideas to promote or radically reject, practicing inquiry, co-research and self-education, without falling into a rhetoric of knowledge disconnected from the desire for a dignified life and to move beyond the concepts of public or private in order to pursue the common.

THE PARADIGM OF DECENTRALIZATION

"Do-ocracy": autonomous commons and capitalist integration[1]

Social political and economic implications

The Great Recession of the first two decades of the 21st century is characterized as a systemic crisis caused by deliquescence of the political, economic and social infrastructures that structured the industrial age, with no new equilibrium having yet been established. In this context, it seems natural that a desire for new freedoms is developing, something only possible if new institutions are built. Nomadic subjectivities prefer "doing" rather than launching themselves headlong into frustrating ideological battles and they wish to experiment with new forms of organization that link the creation of income to ways of life that don't conform to today's normativity.

This new autonomy is making more and more attempts for creating some kind of commons and forms of cooperative, ecological and post-capitalist production: these attempts are often organized starting from the local and extending to the global (glocal), born from a self-managed egalitarian ethic. There are concrete experiences of like the Integral Cooperative like, in Europe, the Integral Cooperative of Catalonia (or *CIC*), which reflects one of the first and perhaps most advanced examples.[2] Autonomous commons politically qualify the

---

1 This paragraph and the next two parts of the article contain "Unblock the chain"(Griziotti, 2014).

2 "The Cooperative Integral is a tool to create a counter-power from below, founded

all-inclusive collaborative commons, a very general and non-political term used by Rifkin and others. The sharing economy, social networks, the internet of things and the growing multitude of "alternative" initiatives that come in the form of startups, associations, coops, etc. that navigate the swamp of the Great Recession, are all in fact part of the collaborative commons.

The prophetic "experts" tell us that this generic cooperation, roughly horizontal, will replace the current hierarchical system, causing its inevitable decline. This assumption is belied by the fact that cognitive capitalism is making the sharing economy its battle horse. It is, in fact, the main "product" offered by companies based on large platforms built to 'suck' income from prosumer network labor. It is a trend that, in addition to the traditional companies like Google, Apple, etc., the two most noted and recent examples being Uber, which makes us all become precarious, hated and exploited taxi drivers, and Airbnb, that transforms our houses into hotels.

We will try to put another arrow in our quiver with the hypothesis that this capture of value constitutes something vital to which we referred to previously. A submission which an ever more extensive autonomous cooperation opposes in a confused but widespread way. Compared with the previous era of identity and belonging, where a clear dichotomy between capital and the working class existed in production, today it is difficult to see such a clear separation, besides the one that divides the multitude of mostly precarious cognitive workers and the narrow financial oligarchy, the famous 1% of the Occupy slogan.

Similarly, there are no clear boundaries nor continuity over a wide spectrum that ranges from autonomous experiences that, as well as claiming a non-capitalist ethic who refuse any compromise, to those that are more or less subsidized, remote-controlled or infiltrated by supporters of the liberal economy, whether public or private. In France, for example, institutions are increasingly present in these areas

---

on self-management, self-organization and direct democracy, for aiu- tare to overcome the current state of subordination, and dependence of the systematic structures but to advocate and practice a scenario of freedom, full awareness, free of authority and where everyone can self-determine under equal conditions. The Cooperativa Integral Catalana (CIC) was created in 2010 and currently has 600 productive projects, 30 social centers and 5,000 people involved." (Duran, 2015).

where they introduce ICT agencies that have the specific task of promoting and subsidizing the creation of startups. This activity, however, only favors the few techno-privileged, leaving the vast majority struggling with unemployment, underemployment and precarious labor conditions.

Even if we observe the same magma from the opposite point of view, the situation is no different. It is difficult and perhaps misleading to look for classifications or specific roles amongst *hierarchical* multinationals, those that are *netarchical* (Bauwens & Kostakis, Network Society and Future Scenarios, 2013, 36) like Airbnb, Google or Facebook and those that are *distributed*, like "bitcoin and the realization of the anarcho-capitalist and liberal dream in which every human being is an entrepreneur in search of its own profits" (Bauwens, 2013). All these trends blend and operate over a continuum of productions and, on one side, intensify the pace of increasingly precarious wage labor exploitation and, on the other, create income through the free labor of the prosumer.

In a review of the Ippolita anarchist collective's *In the aquarium of Facebook* (Hippolyta, 2012), Benedetto Vecchi (Vecchi, 2012) sketched a preliminary diagnosis of osmosis between the blood enemies of two opposing anarchies: leftwing anarchy and capitalist anarchy. According to the Vecchi, the pamphlet is a profound but incomplete attack against Facebook and anarcho-capitalists in general. Two fundamental points are missing: the structural constraints that bind capitalism to oppressive state power are not clear, and there is no trace of any critique of the contemporary social production relation that vampirizes free labor, just like Facebook does with its users.

This lack of clear criticism against anarcho-capitalized highlighted by Vecchi is revealing in terms of what happens with collaborative commons. There are at least a handful of elements that allows us to understand the difficulty in establishing clear demarcation lines in the strategic issue of modes of organization. The first is methodological: although neoliberal practice is ubiquitous in today's society, Generation Y prefers action. In Noisebridge, the historical hackerspace in San Francisco, we find a concept of action-based democracy, or *Do-ocracy*. Here it may not be appropriate to investigate whether this propensity for "doing" before reflection or discussion is a consequence of certain (ideological) excesses found in previous generations, the collapse of "real socialism," constraints resulting from the pervasive cynicism of more than forty years of unbridled capitalist competition or, most likely, a mixture of these causes.

The second element is the decentralization permitted by increasingly sophisticated P2P technologies. This is the common ground shared by these two anarchies, where the same technology can be implemented to build two politically opposite mediations. A third essential element to consider is the relationship between decentralization and state of the art network technologies against the backdrop of widespread (techno)logical intelligence.

Unblock the chain: the blockchain as algorithmic dystopia and tool of the common

Satoshi Nakamoto, the pseudonym for Bitcoin's Silicon Valley inventors, presents the socio-political consequences P2P technologies:

> [Bitcoin is] very attractive from the point of view of a libertarian if we knew how to explain it clearly. I, however, am more comfortable with code than with the words…
>
> …we can win a great battle in the arms race and conquer new territories of freedom for several years. Governments have the ability to cut the heads of centralized networks like Napster, but the networks purely in P2P networks such as Gnutella or Tor they seem to be able to save (the head).[3]

In fact, Bitcoin – and almost all cryptocurrencies – are not only attractive to libertarians.

Like with Facebook, an aquarium also exists around Bitcoin that is functional to financial oligarchies. The individual participant trying to leave the Bitcoin enclosure is forced to pass under the yokes of financial control and levy, for example, when exchanging cryptocurrency into Euros or Dollars. Naturally, for the oligarchy – operating legally or illegally – these rules do not apply. But, in this game, Bitcoin

3 *Nakamoto* means that in applications where there is no central website which however works according to the P2P mode (see glossary), repression cannot technically target and prosecute the alleged perpetrators.

doesn't amount to much when compared to the tax havens and classic money laundering practiced by the World Bank or through "legal" means in, for example, Luxembourg, ruled for eighteen years by the current president of the European Commission, Jean-Claude Juncker.

In any case, Bitcoin introduces an element that is proving to be important in the development of P2P technology. The concept of "blockchain" is derived from the accounting principle of an electronic master book in which all transactions of every bitcoin, from the moment it is created, are written. It is therefore a file that is replicated by every user at the beginning of their activity with that particular coin. By participating in the collective effort to keep the blockchain updated, each user-miner provides, among other things, the proof of work that is the incentive to mine new bitcoins. This mechanism is what, in fact, allows a completely decentralized system to not have any central guarantor to maintain security, consistency and integrity.[4]

These ongoing developments extend the capabilities of this principle, including a programming language that integrates code that can execute procedures of any complexity into the blockchain. A new paradigm of applications and digital contracts that are pure P2P emerges, one that is not, therefore, controlled by any central entity. There are examples that are often cited: dynamic electronic voting, file-sharing networks free from copyright and, in general, illegal transactions that fly below the radar of existing fiscal institutions, i.e. financial derivative transactions. In addition, we can add gambling applications, identity attribution programs, universal income... all the way to the first attempts to create Decentralized Autonomous Organizations (DAO), or even of Decentralized Autonomous Companies (DAC).

This is how Vitalik Buterin, class of 1994, co-founder of *Bitcoin Magazine* and more recently *Ethereum*, a platform for creating decentralized applications, outlines DAO/DAC philosophy:

> The software is becoming the most important component of our present world, but so far the research in this field has focused on two areas: First is the AI, which is the software that works exclusively for its behalf, and the second [is made by] tools or software applications

4 For more information on Bitcoin, see: *Bitcoin, the end of the taboos currency.* (Roio, 2013).

> used by man. The question is, is there something in between? If there is, the idea of a software that directs humans, a decentralized enterprise, is exactly that. Contrary to fears, this is not an evil, heartless robot. It is imposed with an iron fist on humanity; indeed, the tasks that such an enterprise will outsource are precisely the ones that require the greatest creativity and human freedom (Buterin, 2013).

Can we not then imagine financial power conceiving of digital autonomous corporations where "peers" are shareholders that extract profit and whose governance is determined by algorithms that guide human labor? Will Blockchain 2.0 be responsible for "activating" zero-hour contracts? Or will collective intelligence be able to create a context where these technologies amplify and enhance the forms of non-capitalist governance to the point of making them unstoppable?[5]

The first attempts to integrate cryptocurrency into autonomous cooperatives are underway. This is the case of Faircoin, adopted by the recent Faircoop, which we will discuss in the coming paragraphs. Besides technically correcting the problem of the enormous, futile and discriminatory energy expenditure required by Bitcoin and the like,[6] Faircoin is just one part of a complex, global project.

The digital commons and technologies of decentralization

The centrality of knowledge in contemporary economics is primarily based on internet technologies and the worldwide web. We have already covered many aspects of the internet and the Web but, last but not least, here we will address its political aspects. We earlier mentioned Facebook's ambition to create an application so well-articulated and enveloping, it would urge users not to use any other parts of

5 See in this regard, the political fictional reflections on "software agents communists that would allow, among other things, effective economic planning: (Dyer-Witheford, 2013).

6 Faircoin introduces an algorithm of mining the principle of proof-of-stake, a kind of presence test by reducing the role of the proof-of-work system and computing power as extraction engine. See also https://fair.coop/ en /? s = faircoin.

the web, as if we bought a tablet or a PC with only one extended and sprawling application.

In the era before the internet and the Web, online services for the public already existed. The most famous example is perhaps the French project created by Minitel (the name of the terminal used),[7] and decided upon in 1978, after the publication of the famous Nora-Minc report on the digitalization of society and the "telematic revolution" (a term coined in this very report). The basic idea of the project was to give each telephone user an electronic directory to replace the paper phonebooks, freely distributed by France Telecom (a public company at the time), in a terminal with a keyboard and video: the Minitel. The terminal connected via the telephone network to a Télétel computer service based on the Videotex norm. Consultation was free, but there were also a number of paid commercial services. Among these, the evocative "Pink Minitel" service was a surprising success.

The Minitel was a great success in the '80s and '90s: millions of terminals were distributed, making France the most advanced country in terms of network connections. The system, however, was based on norms few countries adopted (Videotex), was relatively slow and possessed limited graphics. It was also a "closed" system that did not easily connect to different kinds of terminals. From a political point of view, its strength and its weakness were in the fact that it was a well-conceived project, financed and managed by one centralizing state (France) that, in fact, allowed rapid and widespread distribution, as well as the accessibility of the Télétel service. However, the administration's rigidity and inability to change and develop in order to make it compatible with emerging international standards (Internet, TCP-IP) was fatal. Despite surviving until 2012, it was outclassed by the development of the internet. Precisely because it had been at the avant-garde with the Minitel project, France adopted the internet later than many other European countries.

Another example was the service offered by America Online originally called Q-Link. In 1985, Q-Link, like its competitor Compuserve, was able to offer users of Commodore 64s and other

---

7 The name is derived from Minitel: Médium interactif par numérisation d'information téléphonique and refers to a terminal type to connect to the service French Teletel (standard Videotex) exploited commercially in France between 1980 and 2012.

personal computers an articulated series of online functions like email, chat, news and games, all with proprietary protocols. This meant, among other things, that only subscribers could communicate with each other, and that it was not possible to interact with subscribers on other private services.

This introduction highlights that, for their characteristics, the internet and the web are common goods, fundamentally important tools in today's technological mediation. They are not "products" made by private companies but free services (except internet access) that allow interoperability of devices, applications and networks without limits. Tim Berners-Lee and his collaborators should be recognized for having guaranteed the fundamental qualities of Open Standard* for all the main components of the Web, even in the W3C.[8] In the beginning, nothing assured these qualities, especially if we take into account the pre-existing networks and private, proprietary services. From this point of view, the internet and the web are part of a common, like free software, and even today they remain so, even though they are threatened. With these threats coming from all sides, we will only mention those concerning net neutrality and the attempts to build a two-speed internet, specifically those from multinational telecom companies.

In early 2015, the American Federal Communications Commission (FCC) voted in favor of net neutrality but appeals were immediately deposited by telecom operators against this decision. These same uncertainties remain after the European Parliament's vote on Telecom Single Market regulations in October 2015, which opened wide margins for the creation of privileged services and the end of neutrality. Let us now return to the underlying trend of decentralization and to collaborative P2P applications that are a fundamental part of their technology infrastructure. As can be deduced from the previous paragraph, these applications cannot, in themselves, be guaranteed as tools for the common. In all its variants, cognitive capitalism directly uses P2P technologies since they can be perfectly suited to the process of putting life itself to work. The DAC[9] philosophy, based on the principle of a program where algorithms manage humans, could amplify this trend.

---

8 World Wide Web Consortium, known as W3C, is an organization international NGO that aims to develop all potential of the world wide web.

9 Decentralized Autonomous Companies

One of the causes that contributed to the fall of the Soviet regime was probably due to the inability to create, after the October Revolution, non-capitalist cooperation despite adopting the rules of industrial capitalism. Current trends in autonomous cooperation run the opposite risk, i.e. not transforming itself into political expression capable of changing the present. Despite their relative electoral success, political parties like Syriza in Greece and Podemos in Spain have difficulties in reactivating and strengthening the movement to the point of starting a post-capitalist glocal process; without the social struggles that created them, these parties risk being isolated by financialized governance, and suffer heavy defeats.

The claim for a Universal Basic Income (UBI*), which would enormously enhance autonomous cooperation, is certainly an important axis that these convergences must operate on in order to accelerate the process of political maturation. How can a link between the collaborative processes of P2P and constituent political expression able to create a common system of social relations be built? This is what we will now try to explore.

Loosely coupled movements. Towards a flexible and autonomous post-capitalist organization

In computer design, correlation or dependence means the degree to which each component of a program relies on the others. Low correlation systems (loosely coupled) are those in which each part has, or makes use of little knowledge of the other, separate components; this is an effective and widely adopted mode.

From a social, political and economic point of view, much has been written on the horizontality of contemporary movements, both in a positive and negative sense, with the result that now it seems necessary to analyze what these modes of collective behavior really mean. On the one hand, the relationships within the movement have developed like an internet infrastructure in the sense that the autonomous entities and initiatives that constitute it are connected in a flexible manner and represent a kind of diversified "network of networks." This is a diversification that includes all the complexity of society with political, social, cultural and economic activities. Individual participants, who often hold multiple affiliations in different spheres of interests like collective entities, associations, cooperatives etc., interact, crisscross, unite and separate.

That the current movements, unlike those of the '70s or '80s, prioritize action is also a logical consequence of these dynamics. For example, the belief that any activity, from artistic expression to the search for income, can be politically expressive is widespread. "Doing" is immediately political, while the political ruling class, often made up of the worst national biotypes, is held in contempt, even if this sentiment sometimes takes on populist accents that, if not merely indifferent, feed the radical right. There are the "intermediate" positions of those who believe that the change may materialize through a social compromise between the public and the common, as is being experienced here and locally. Once again, a poignant example comes from Barcelona where the movement against evictions has led to more ambitious social housing programs. The convergence of the PAH committee[10] with communities and local public structures was the decisive factor of the recent municipal electoral success.

"Doing politics" also includes the launch of the global Faircoop project, an initiative created as part of the integral cooperative (in particular of the CIC[11]) and the P2P Foundation. Faircoop aims to federate a broad global set of autonomous Commons autonomy that lean towards "a new world economic system based on cooperation, ethics, solidarity and justice in economic relations."[12] On a completely

10 Plataforma d'Afectats for Hipoteca , Platform of Victims of mortgages whose spokesperson and co-founder Colau Ada is now mayor of Barcelona (6/2015).

11 Cooperative Integral Catalan.

12 16 "September 17, 2014 the Faircoop project was launched, presented as an "Open cooperative global, another a step forward in expanding the integral revolution. "FairCoop is an umbrella project, a decentralized meeting point (self-organization of the Internet, so beyond the control of the central banks and legislation of nation-states), the global scope, which does not pretend to overlook that you are creating and practicing in local areas, it puts in common the tools and knowledge generated at bioregional level in order to exploit them for other related projects in different places (not replace, overlap or cancel; but share, enhance and, for the collective, re-use). " "His goal is to contribute in enabling the transition to a world again reducing as much as possible the economic and social disparities between human beings and at the same time helping to ensure new global wealth, accessible to all humanity in the form of cooperation. "Initially the creative idea of FairCoop was co-launched by Enric Duran (co- founder of the Cooperativa Integral Catalana -CIC-, collaborator of Fondation P2P and Wallet Dark). He set up a development plan for FairCoin (current

different plane, another example is represented by a movement that promotes permaculture.

In this regard Alfonso Giuliani in *Managing the commons in the knowledge economy* (Vercellone et. al., 2015, 124) evokes the "Transition Town" movement:

> The Transition Movement [...] consists of Community Initiatives, which try to transform society into resilient communities organized according to maker logic to cope with environmental challenges related to climate-change, with limited resources and with mutations in the world of work brought about by economic crisis. One of the main characteristics of the Transition Movement is to believe that all these problems can be addressed through co-production and community collaboration. No coincidence that the two fundamental principles of the movement are: a) individuals have immense amount of creativity, ingenuity and ability; b) if they were acting as individuals, communities would be able to create a way of life significantly connected, more vibrant and more satisfying than what we live.

The fluid that flows in the networks as digital bodies is living matter, a hybrid of nature and technology. A sap that, under certain conditions, escapes to the sensors and algorithms that are predisposed to control it, creating tumultuous growth. Local authorities respond by instituting laws and regulations that legalize the use of algorithms and sensors for the general surveillance of networks and communications, undermining more and more the Rule of Law. In this regard, France's scandalous "Personal Information Law," presented by a social-liberal government and recently approved in France, is an illuminating

---

reserve currency of FairCoop value created by criptomoneta FairCoop people external to the project and then fallen into disuse). This plan is shared with members of the P2P Foundation (such as Deadlift Troncoso and Michel Bauwens) of CIC and Dark Wallet (as Amir Taaki and Pablo Martin), and leads to a situation where other people and projects can act in various countries, such as cooperating and working together on the net, and form the promoter team that creates and launches the first FairCoop proposal of the project last September." (Faircoop, 2015).

example.[13]It is law *à la* "Patriot Act," which, after the *Charlie Hebdo* attacks in Paris, legalises mass network surveillance without any judicial oversight. This situation dramatically worsens with the promulgated long-term emergency and constitutional changes in France as a result of the terrible attacks of November 2015 in Paris. These examples are nothing more than a further, dramatic evolution of the clash between two necropolitics of which we are all collateral victims.

This control is exercised in multiple ways: innocent-looking programs and algorithms, like real time analyses carried out Twitter,[14] allow us to understand instantly the generalized mood on any topical subject and can become tools for emotional manipulation with political ends. In this hybrid of technique and nature in collective behavior, the dynamics of biopower and of General Intellect intersect and sometimes clash. The former pushes vital subsumption to paroxysms that make us fear the possible extinction of our species. The latter scours the common with an instinct for post-capitalist survival.

On the other hand, social movement dynamics appear to mimic certain behaviors of living structures. The mycelium, for example, is a mixture, a natural network that lives and branches mainly in the woods and forests but, growing across the ground, is invisible to the naked eye. In suitable conditions – temperature, humidity and food availability – its growth becomes tumultuous, at rates of almost one centimeter per hour and, in just a few hours, mycelium can suddenly, without any prior or predictable signals, produce large amounts of fruiting bodies, commonly called mushrooms.

Perhaps we could say that the humus and the ways in which waves of social movements develop are especially technological and intangible assets, taking us back to the biohypermedia structures we have addressed at length. However, the sudden and unforeseeable ways in which these explosions occur share strong similarities with what happens in nature.

---

13 17 "Loi sur le renseignement personnel" No. 2015-912 was promulgated in France July 24, 2015.

14 For example, the progress monitoring of influenza epidemics is done by analyzing the Internet flows that geolocalize the recurrence of keywords like "flu" or "fever." See the private commercial service analysis. It intended to "digital communication" apparently used by professionals Services of the French Republic Presidency: http://www.visibrain.com/fr/.

THE COLLAPSE OF BALIVERNA[1]

"We are already after" because we hear louder and louder crunches coming from the Baliverna where you're imprisoned.

"We are already after" because you have boarded yet another Bismark battleship destined to sink but we'll try to make it a Potëmkin of the common.

"We are already after" because your mainstream media is in decline and we no longer believe your story telling.

"We are already after" because the only way to continue the human story is the construction of the common.

"We are already after" because science tells us that even the cyborg can rise up and win.

"We are already after" because the network of networks is the model of the common.

"We are already after" because we know how to become machines and nature.

"We are already after" because we need no gods nor transcendence to become machines and nature.

"We are already after" because, like André sang to the revolt, even if you believe you're acquitted, you are forever involved.

---

1 A Story by Dino Buzzati, 1954.

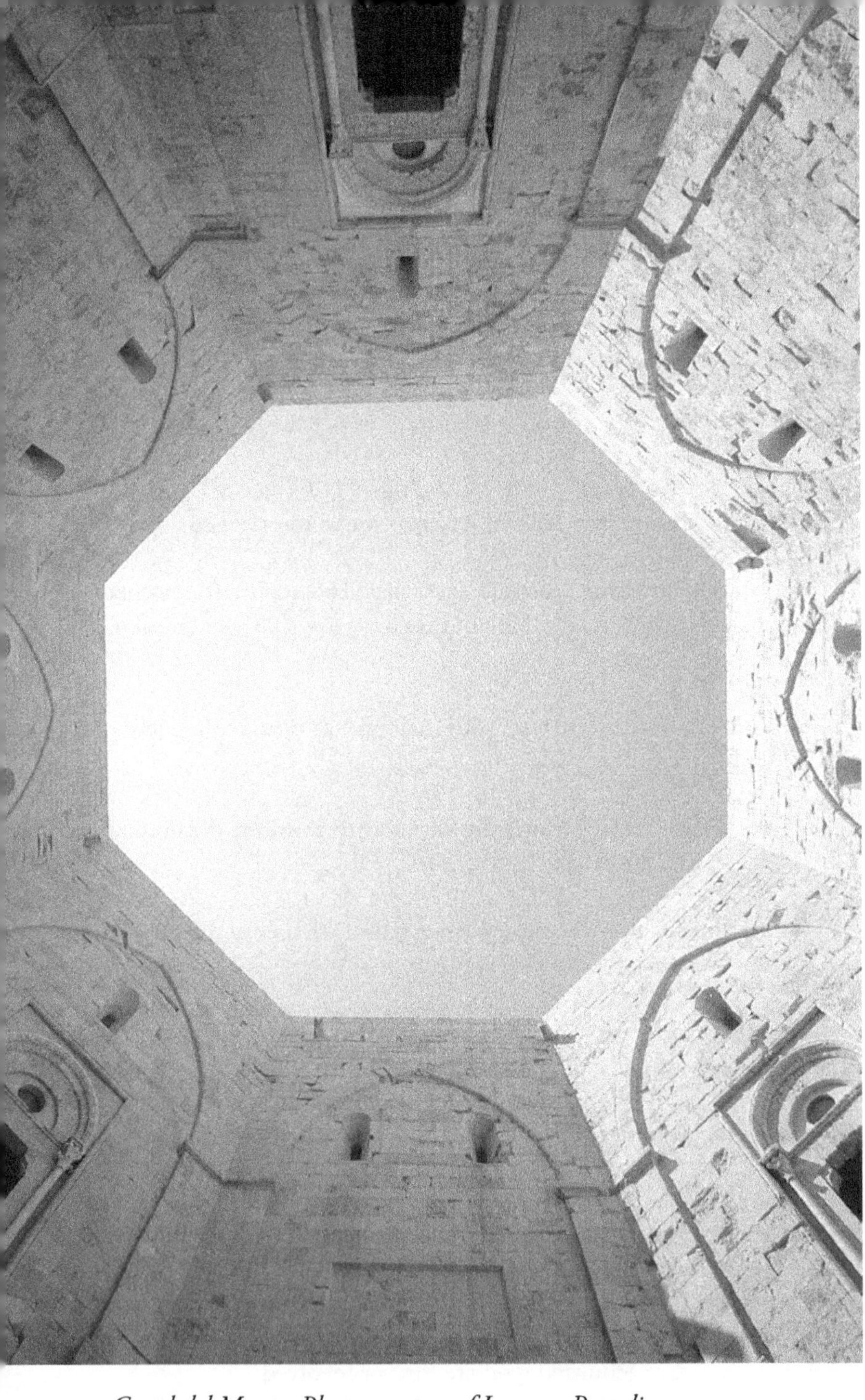

Castel del Monte. Photo courtesy of Lorenzo Papadia.

FINAL EVASION:

The Byte of Castel Del Monte and the Wonders of the Common

Castel del Monte, located in the Municipality of Andria in Apulia, is one of the most symbolic and fascinating monuments of the Middle Ages. It was built in the 13th century by Frederick II of Swabia, a monarch that so deeply marked his era as to be nicknamed "Stupor mundi." Among other things, he spoke six languages and had Greek, Latin, Arabic and Hebrew courts, meeting places for cultures and technologies. This is the context in which the construction, called by some a stone book, was developed.

To this day, immersion into Castel del Monte provokes strong sensations. First, the chosen site where it was built about eight centuries ago – an isolated hill overlooking the vast karst plateau of the High Murge – is breathtaking. Perhaps, to paraphrase Simondon, one could imagine it being a key point of the primitive magical universe. The building rises in the most absolute solitude and only through a single, north-facing mullioned window can one see the city of Andria, loyal to Frederick II, in the distance.

At first glance, the most striking aspects are the complex and harmonious geometry of the forms and its octal obsession: the construction has the form of an octagon and, in the eight corners, just as many octagonal towers rise; the courtyard is octagonal, with eight rooms on the ground floor and eight upstairs, all trapezoids, all equal to one other. A refined repetition of eight, a number that recalls the symbol for infinity[1] more than any other, characterizes this unique monument in style, form and content.

And it is precisely from this characteristic that we find the link with our digital and algorithmic world. Octal numbers, along with binaries and hexadecimals, are the basis for digital calculation and are part of the network's very DNA. Computers have always pivoted around the powers of two. The first miniaturized processors that allowed the development of personal information technology – the Intel 8086 and 8088 and the Motorola 68000 and ZX80 – were built around 8-bit registers capable of managing octal numbers. Today, we have arrived at 64-bit processors which is, once again, a power of eight, the number which establishes the amount of compromises between binary code and the bios, between two hemispheres and ten fingers. In the movie *Avatar*, Pandora residents have eight fingers and count in octal. Some people use octal numbers to count: "the Yuki tribe, a native people of California, and the people Pamean in Messenia physically possess a counting method using the space between the fingers, rather that the same fingers to count."[2]

The strength and charm of Castel del Monte expresses can only be the result of an exceptional common, the capabilities of cultural production of that era "both from antiquity as well as the Northern European Cistercian tradition, from the Castles of the Omayyadi Desert and the fortified monasteries of the Middle East and North Africa.[3] Another point that binds us to this work is the use of algorithms.

The castle is the result of amazing astronomical and trigonometrical research, so during equinoxes or other specific days, the play of light creates special effects and while the spatial distribution of the main architectural elements obeys precise instructions issued by the Sun throughout the year. Eight centuries later, its function remains enigmatic, making it even more fascinating because the criterion of the user experience from today's marketing is not applicable. There

1 Rotated 90 degrees, the number eight becomes the symbol of infinity.

2 From https://en.wikipedia.org/wiki/Decimal. Translation. See also: "The typology of Pame number systems and the limits of Mesoamerica as a linguistic area "(Avelino, 2006).

3 In http://whc.unesco.org/fr/list/398. Translation author.

is no purpose of power and, despite the name it bears, is not a military construction, nor was it used for entertainment since it was not intended for hunting, as supposed by experts and researchers, and it does not seem a suitable place for a long stay.

Castel del Monte seems a direct contrast to any compulsive or finalized use of technology. The purity and beauty, emerging through techniques that do not convey any teleology, would today seem inconceivable. Castel del Monte, like other works from the era of Frederick II, is the fruit of a multi-ethnic and multicultural common that makes us wonder if the ugliness of so many of today's "great works" are nothing but the result of a corrupt common.

This monument's apparent lack of purpose integrates with other extraordinary characteristics enough to allow us to use it as a metaphor for the construction of the common in network society. It represents an immense challenge that we can face because we have powerful technological tools, but this alone is not enough. Humans remain responsible for the spark.

ACKNOWLEDGEMENTS

BASED ON THE SUGGESTIONS OF FRIENDS WHO SUPPORTED ME IN writing this book, in the introduction I tried to give the subjective aspects that would provide the reader with a more complete framework for my motivations; now is the time to thank them. It seems fair to me to start with Giorgio Bonazzi a.k.a. Mingus, friend and coach, as they now say. It all began one evening a few years ago in Milan on Via degli Scipioni, when, after reading one of my first articles, Mingus said "if you decide to write a book, I know the publisher who will publish it," giving me the initial spark. Laura Toppan has assumed with great generosity, friendship, seriousness and professionalism the challenging role of "compensating for the linguistic and literary weakness of an engineer who emigrated decades ago," dedicating considerable energies to my words with detailed and expertly annotated readings. Carlo Vercellone is my dear friend who patiently and tenaciously guided and helped me in the most difficult conceptual passages and, in particular, on the aspects of political economy concerning cognitive capitalism, an important concept of which I think he is one of the greatest, if not the principle, contributors.

A special thanks to Martino Saccani, a young conceptual artist of great talent who has done me the honor of creating the beautiful cover. Thanks also to those who provided images including Valentino Bellini (in the E-waste evasion) and Lorenzo Papadia (in the Castel del monte evasion). All other images are in in Creative Commons.

I'd like to thank my group of friends and companions who read the manuscript with a critical eye. Each, through his or her own skills and sensitivity, has contributed to adding another point of view, encouraging me or improving the result: Alfonso Giuliani for his pertinent economic notes and on other technical aspects, and for having drawn my attention to the need for a certain formal rigor; Gaetano Grasso for his advice and valuable suggestions on the Italian language; Massimo Meloni, above all for his fine criticism of the political situations explored; Saro Romeo for his enthusiastic encouragement and suggestions on the first section.

I'd also like to thank Isabella Annesi-Maesano and Gianfranco Pancino, dear friends and researchers with great experience and expertise in the biological and medical field for helping me and providing valuable explanations and information on the aspects related to biogenetics, allergies and autoimmune diseases. I owe a debt of gratitude to Gianluca Giannelli for the collaboration in the article on e-waste 2.0 and the intense discussions over the summer of 2013 in the Parisian parks and along the banks of the Seine, which inspired the chapter on "Belonging and crossings;" another to Valentino Bellini, a photographer who fights e-waste with his courageous Bit Rot Project, for not only the photo, but also for his knowledge and for the urge to denounce this havoc.

A thank you to Nanni Balestrini who helped me build confidence by publishing me in *Alfabeta2* magazine, as well as for always encouraging my initiatives. To Toni Negri, a long-time friend, I owe many of the teachings and many of the concepts used in the book. Pantaleo Elicio and Jason Francis Mc Gimsey are part of the small circle of first responders who have advised, helped and supported me in various ways, all useful; I have also taken advantage of the sympathy of many friends with prodigal attention and suggestions, including Giuliano Righi Riva, Latino Imparato and Sylvain Quendez.

In addition to their patience in putting up with me, I owe my family essential contributions. My companion Tiziana, my first reader, supporter and council on the most important points, also ran the race to correct typos and mistakes, and my son Diogo taught me everything I know about the contemporary video games that he intelligently analyzed for me. Last but not least a heartfelt thank you to my dear friend Tiziana Terranova, who helped me, criticized me and encouraged me, and who was able to intensely grasp and synthesize the spirit and facets that I tried to convey throughout the book in a preface that touched me deeply. Having often dealt with space-time, I'd like to close with an acknowledgement of two places: The Swedish Cultural Institute in the Marais in Paris which, in summertime, offers the magnificent garden of the Hôtel de Marle, a place of great harmony that stimulates concentration; and the starred high-ceilinged open-space of San Cassiano di Lecce where, on an ancient carpenter's table, *maker ante litteram*, I am writing these last lines.

As far as time is concerned, I have no doubt: the night is my favorite.

GLOSSARY

Anthropocene/Postanthropocene

The anthropocene is "a proposed epoch dating from the commencement of significant human impact on the Earth's geology and ecosystems, including, but not limited to, anthropogenic climate change. [...] The term was widely popularized in 2000 by atmospheric chemist Paul J. Crutzen, who regards the influence of human behavior on Earth's atmosphere in recent centuries as so significant as to constitute a new geological epoch. [...] It is a combination of *anthropo-* meaning 'human' and *-cene* from *kainos* meaning 'new' or 'recent.'"

Although an increasing number of scientists use the term, the debate continues on the relevance from the generic term anthropocene as a geological epoch within the geological time scale. Jason Moore expresses this criticism and introduces with others, like Christophe Bonneuil, the term capitalocene:

> Are we really living in the *Anthropocene*, with its return to a curiously Eurocentric vista of humanity, and its reliance on well-worn notions of resource- and technological-determinism? Or are we living in the *Capitalocene*, the historical era shaped by relations privileging the endless accumulation of capital?[1]

With the term postanthropocene, then, a geological era to come where machines (or possibly a posthuman civilization), which have assumed their degree of autonomy, contribute to structural changes in the climate is designated.

1 https://www.versobooks.com/blogs/2360-jason-w-moore-anthropocene-or-capitalocene

App

App is an abbreviation for "application." An app is a synonym for a temporary application program that works on the Internet, on a computer or on any other device such as smartphones, tablets, game consoles, etc. Currently, the term primarily refers to smartphone applications. The apps used on mobile devices originate with the iPhone and the birth of the App Store followed then by the Play Store for Android systems etc.. In addition, the world of apps is an alternative to the use of the World Wide Web and browsers, since they are independent of it.

Augmented Reality (AR)/Virtual Reality (VR)

Augmented reality (AR), or computer-mediated reality, is the enrichment of human sensory perception through information, usually electronically manipulated and conveyed, that would not be perceptible with the five senses. The dashboard of the car, the exploration of the city by targeting the smartphone or remote robotic surgery are all examples of augmented reality. Virtual Reality is an interactive and immersive, visual, sound and tactile computer simulation of real or imaginary environments. The purpose of virtual reality is to allow one or more people a sensory-motor and cognitive activity in an artificial world, digitally created, that can be imaginary, symbolic or represent a simulation of certain aspects of the real world.

Batch

The term refers to the execution of a series of programs (jobs) without operator intervention. It was used in the punch card programming period. In that context, programmers usually did not have direct access to the computer, but prepared their own programs "off-line" and passed them on to a system administrator who then had the task of keeping them running whenever possible (queuing them after other programs running and often merging several programs into a single execution unit), then returning the processing results of the processing to those concerned.

Biohypermedia

For Giorgio Agamben "questions of terminology are important in philosophy" (2006, 5) and "terminology is the poetic moment of thought." The dynamics of network flows tend to reproduce the socio-economic relations and the tensions of external reality in a sphere. The more articulated this sphere, the more intense the internal-external exchange is. The result is a new globality in which network contents and the physical world intertwine and integrate, and this new entity deserves to be designated. A brief etymological review of the terms created to name to this new globalizing entity allows us to introduce some aspects that characterize it; be they political, technological, social, cultural or spatio-temporal.

Marx's General Intellect, belonging to the famous Fragment of Machines in the *Gundrisse* (2012, 478), is probably the first intuition of what would lead to the current transformations.[2] Although this concept does not seem to directly integrate the notions of space and time, it indicates, according to Marx, a new dimension, in which collective knowledge becomes a force of realization that has an impact on the division of labour and technological progress.[3]

The noosphere is a term coined by a celebrated Soviet scientist Vladimir Verdanski and a French paleontologist and Jesuit philosopher, Pierre Teilhard de Chardin. The root "nous" comes from Greek philosophy and roughly means "cosmic intellect." Verdansky juxtaposes the Noosphere, which could be translated into the sphere of universal knowledge, with the geosphere, that is to say to inanimate objects and to the biosphere, the sphere of the living.

Marshall McLuhan, coining the concept of Global Village (1991), is among the first to mention a virtual world modeled by electronic media. Anticipating the arrival of the Internet by about thirty years, McLuhan refers above all to the new electrical media of his time, including television, which call into question the supremacy of the written word.

Rejecting the concepts of biosphere and noosphere, considered the industrial era as that of insects, Deleuze and Guattari speak of a

---

2 See also Vercellone, C., 2007

3 "Marx announces the arrival, after the phases of the formal and real subsumption of capital, of a new phase of development in the division of labor. It is here that Marx talks about 'General Intellect' to characterize the impact of this change in the division of labor and in technological progress." (Vercellone C., 2007) [our translation]

"*mécanosphère*"[4] where "the apparent order can be reversed" and technological and cultural phenomena become breeding stock for insects, bacteria and microbes. One could then deduce that the present cognitive era would be characterized by viruses, entities at the frontier of the living...

The notion of "cyberspace" coined by William Gibson, in *Neuromancer* (1986), is the most widely diffused, perhaps, precisely because, as the writer himself explains,[5] it spontaneously emerged, without previous rationing. He refers to *a consensual hallucination experienced daily by billions of legitimate operators*, a parallel, virtual world that tends to overlap rather than merge with the real world. Gibson, in fact, contrasts it with the "meatsphere" which, for him, is the world of flesh and bodies. The same, but even more markedly, is the term info sphere, commercially recovered by IBM as opposed to the biosphere.

Definitions such as 'information society,' 'knowledge society,' 'network society,' the latter often used by Manuel Castells, have a strong sociological and cultural connotation and deny the influence of the network on behavior and knowledge. None of these terms emphasizes the fact that, with the pervasiveness of the networks, hybrid devices are integrated into all spheres of society: the natural, cultural, political, affective and economic all consequently include both the infosphere and the biosphere. It is a global biopolitical and biocognitive machine. If, on the one hand, the multitudes can free themselves from subdued modes of production, they may render obsolete many of the cooperative functions of capitalism; on the other hand, it is also biopower's preferred instrument in that it is able to exercise unlimited, capillary and neuro-pervasive control.

We have defined this as a Biohypermedia machine because it extends the power of hypertext and hypermedia, developed with the web, to all spheres of life. Biohypermedia elevates to a fourth power

---

4 G. Deleuze, F. Guattari, *A Thousand Plateaus*.

5 "All I know about the term "cyberspace" when I coined it was that it actually resembled a fashionable term. It seemed evocative and essentially meaningless. It was indicative of something, but it had no true semantic meaning, even for me, since I saw it emerge while I was writing it on the page."

Gibson's commentary on the origin of the term "cyberspace" in the documentary *No Maps for These Territories*, directed by Mark Neale, 2000.

the previous paradigmatic jumps of modern media, arbitrarily taking as its starting point the printing press born with the Gutenberg revolution (McLuhan, 1991). The second power is represented by the mass media of the industrial revolution, that is: newspapers, telephone, radio, television, etc.

The third is the hypertext intuited before the Internet and the web by artists and writers[6] such as Julio Cortàzar who, in 1963 with his Hopscotch invents a manual hypertext, giving indications on how to read his novel in more than one way; or Jorge Luis Borges who describes the hypertext in his short story "The Garden of the Bifurcated Paths"; or again, Italo Calvino with his novel *If a Traveler During the Winter Night*; finally, there is Georges Perec in *Life a User's Manual*, 1984). These works were all written in a way that created new dimensions through reading variables and structures strongly interwoven with their contents.

The hypermedia that extends the principle of hyperlinks to all media has had predecessors such as, for example, in cinema, the now common method of flashbacks or interlocking stories. The entire World Wide Web is a huge, hypermediatic protean entity. Some traditional media, such as large newspapers, have difficulties in finding a model that allows them to exploit the full potential of hypermedia; often on their sites the articles are the simple reproductions of what is printed on paper without including hyperlinks or hypermedia.

Biohypermedia breaks down the physical and logical barriers of pre-existing media. With the development of the "Internet of Things," the possibility for each object to become a biohypermedia device extends without limits and, consequently, the categories of dialogue tend to multiply. In industrial capitalism, humans act on inanimate matter by means of instruments and machines, while today machines become subjects that interact autonomously and, among other things, act on humans, on our biological regulation and on our organs for the control of or rebalancing of our vital functions.

Barriers fall because a considerable part of the interactions of the informational exchange no longer passes through the logical mediation of text or even language, an unprecedented and perhaps

---

6 Italo Calvino evokes the term "hyper novel" in a conference that he gave at Harvard University in 1985, later published posthumously in *American Lessons. Six proposals for the next millennium.*

decisive way of reducing the "fourth discontinuity,"[7] that between man and machines.

Biopower

Biopower is a type of power that is exercised over life: the life of bodies and that of the population. According to Michel Foucault, it gradually replaces the monarchical power of life and death. The exercise of this power is a way of governing humans.

Bitcoin

The term bitcoin (from 'bit': unit of binary information and 'coin': currency) refers both to an Internet-based payment system and to a unit of account used by the payment system. Unlike most traditional currencies, Bitcoin does not use a central entity: it is a P2P and open source computer system, which uses a database distributed between network nodes that keep track of transactions, and uses encryption to manage functional aspects such as the generation of new currency and the attribution of bitcoin properties.

Cloud Computing

Cloud computing is a method of delivering information resources, such as storage, processing or transmission of data, characterized by the availability on demand through the Internet from a set of existing and configurable resources. Cloud computing is characterized by the flexibility of the use of resources and, in some cases, replaces the internal computing centers of companies.

CNC

Computer numerical control machines, which until the eighties were used only for high-precision workings, are today widespread and used in almost every field of mechanics. The technology of CNC machines covers some of the branches of mechanics, the most common being

7 The fourth discontinuity is that between humans and machines according to the historian and philosopher Bruce Mazlish; see Mazlish 1993.

press brakes, punching machines, lathes, milling machines, welding machines and sheet metal cutting machines.

Cognitive capitalism (biocognitive) and the knowledge economy

"Cognitive capitalism means the passage of industrial capitalism to a new phase in which the cognitive and immaterial dimension of labor becomes dominant from the point of view of value creation and the competitiveness of companies. In this context, the central issue at stake in the valorization of capital and the forms of property is the correct appropriation of the common and the transformation of knowledge into a fictitious commodity." (Negri & Vercellone, 2008)

"The term 'knowledge' designates the development of an economy based on the diffusion of knowledge and where the creation of knowledge becomes the main issue at stake in the valorization of capital." (Vercellone, 2002).

Cognitive capitalism is, therefore, the dominant structure while the neoliberalism is the mode of regulation and organization and of the policies of command over the society.

"Compared to the approaches of the knowledge-based economy, the hypothesis of cognitive capitalism strongly reaffirms the capitalist nature of the current process of transformation, which means opposing the very concept of the knowledge-based economy with that of cognitive capitalism, which subsumes it and frames it in a series of institutional forms. We can see this process of subsumption through various mechanisms such as intellectual property rights and new labor control mechanisms which, instead of encouraging the development of a knowledge-based economy, block it with the aim of being able to capture the value and knowledge produced by knowledge and turn it into a capital, a fictitious commodity."[8]

"Biocognitive" is an evolution of the term "cognitive," introduced to indicate cognitive capitalism's influence on life and is translated and used in various articles and essays by Cristina Morini and Andrea Fumagalli. We have used it as a qualification in cases where the links between the cognitive aspect (e.g. of labor) and life (bios) are particularly strong.

---

8 http://www.uninomade.org/capitalismo-e-conoscenza-intervista-a-carlo-vercellone

This evolution also seems to go in the direction indicated by Maurizio Lazzarato, who reproaches the theory of cognitive capitalism for minimizing the production of subjectivity in the knowledge economy. This seems to ignore "machinic enslavement [which] works from a-signifying semiotics (stock exchange indices, the currency, mathematical equations, diagrams, computer language, company and national accounting, etc.) that do not pass through consciousness and representations and do not have as a referent the subject."[9] (Lazzarato, 2014, 17). Consequently, throughout this work we tried to highlight the "a-signifying" semiotics that populates the biohypermedia space (smartphone notifications, socio-economic and now somatic profiling, ranking, filter bubble, etc.).

Common

In this text, we sometimes use the term "common, in the singular, to distinguish, for example, from that of "common goods." The common to which we refer is therefore not an intrinsic characteristic in the nature of goods but is born, it is "produced" through social cooperation and, more generally, human cooperation. It takes on a new dimension in developing the knowledge economy: "The approach of the 'Common in the singular' is inscribed in interpretation of the thesis of cognitive capitalism. [...] This approach considers in particular that it is not in the nature of goods, but in the new historical characteristics of cognitive work that the ontological principle of the new Commons of knowledge must be sought, whether material or immaterial, social or land..." (Vercellone et al, Managing the commons in the knowledge economy, 2015, 38).

Co-research

The co-research method came to light in the early sixties, developed by a collective whose members included Romano Alquati, Toni Negri and Mario Tronti.

> The research concourse that was born from militant research in the field with workers from Fiat Mirafiori and

9 Lazzarato, 2014.

other Piedmonts' factories (Olivetti, Lancia) is, at the same time, an activity of investigation and a process of mutual knowledge and transformation of the identity of the researcher and of what began to be identified as "worker subjectivity" in those years. These investigations were important in the cycle of conflicts inaugurated in Turin in July 1962 in Piazza dello Statuto, the forerunner of the Italian uprisings of '68.

Co-research is an interventional practice which, by placing the militant researcher on the same level as the subject under investigation, eliminates the separate figure of the vanguard so dear to the logic of the Leninist left, and allows the relationship between theory, practice and organization to be reshaped horizontally and circularly.

Research is a social and political relationship that cannot be formalized in a method that allows us to read, even in periods of passivity, the signs of conflict to come, the informal organization and the constitutive ambivalences that lie in the gap between technical composition (objective articulation of the workforce) and political composition of class.

Co-research therefore produces effects at the same time as it is constructed collectively, because it is a space in which the subjectivity of the researchers and the researchers can be expressed. It is therefore an activity that makes it possible to build new possibilities. (Armano & Sacchetto, 2012, 3).

Cyborg

Cyborg is the contraction of cybernetic organisms. A cyborg is a human being - or at least another intelligent living being – who has undergone modifications or has received mechanical, computer, bionic or other grafts. The feminist theorist Donna Haraway, in her "cyborg theory" maintains that the natural tendency of human beings is to reconstruct themselves through technology in order to distinguish themselves from the other biological forms of the planet: a project that starts from the first forms of manipulation of the human body and continues today with the use of technological prostheses and the development of genetic engineering. The desire to improve what determined nature, according to Haraway, lies at the very origins of human culture.

Data farm

In computer science, the term server farm or data farm is used to refer to a series of servers located in a single environment in order to centralize their management, maintenance and security. It is not uncommon, in fact, to find server farms consisting of hundreds, if not thousands, of servers. The realization of a data farm allows to distribute the costs of construction and management of a secure structure and to reduce the costs for connectivity.

Enclosure

This term refers to the "enclosure movement and expulsion of peasants from the communal lands, [...] A fundamental step in the process that leads to the establishment of the principles of absolute private ownership by making land a fictitious commodity, i.e. an asset that can be freely purchased and sold on the market." (Vercellone et al., 2015, 8) The so-called "Enclosure Acts" mainly damaged the peasants, who could no longer reap the benefits derived from those lands, favoring instead the large property owners. Today, the term is used more generally to designate all legal forms, technical or by force, that will limit or exclude the uses of the commons.

Free Software

Free software is software published under the terms of a free license, which encourages its use, study, modification and redistribution. According to the Free Software Foundation (FSF), founded by Richard Stallman, software can only be defined as free if four "fundamental freedoms" are guaranteed:

Freedom 0: Freedom to run the program for any purpose.
Freedom 1: Freedom to study the program and modify it.
Freedom 2: Freedom to redistribute copies of the program in order to help others.

Freedom 3: Freedom to improve the program and publicly distribute improvements so that the entire community benefits.

Access to source code is a condition for exercising freedoms 1 and 3. The FSF further stipulates that:

> Firstly, these freedoms must be irrevocable. Everyone must have the opportunity to enjoy it without the need to prevent anyone. Redistribution of a program must be possible in any form, including supplying a compiled program, provided that the corresponding source code is available. The user must be able to use free software of which he is not the author.

FSF

The Free Software Foundation (FSF), founded by Richard Stallman on October 4, 1985, is responsible for removing restrictions on copying, redistributing, understanding, and modifying computer programs. The FSF operates by promoting the development and use of free software in all areas of computing. The Free Software Foundation is the creator of the GNU General Public License, commonly referred to as the GNU GPL or simply GPL.

GNU GPL

The GNU General Public License, commonly referred as the GNU GPL or simply GPL, is a free software license, originally drafted in 1989 by Richard Stallman to distribute the programs created under the GNU Project of the Free Software Foundation (FSF). The *copyleft* provided by the license made it possible to deploy Linux-based systems (and then countless other applications and programs), giving programmers who contributed to the kernel the confidence that their work would benefit everyone. These systems remain free, and cannot be exploited by software companies that would have offered nothing in return to the community.

ICT

New Information and Communications Technologies (ICT), are the set of methods and technologies that implement the systems of transmission, reception and processing of information (including digital technologies). Sometimes NTICs are also used where N stands for "new."

Interrupt

In computing, an interrupt or interruption is an asynchronous signal that indicates the "need for attention" by a device aimed at a particular service request, a synchronous event that allows the interruption of a process if certain conditions occur (process management) or more generally a particular request to the operating system by a process running. With the improvement of operating systems, more complex mechanisms have been introduced, such as Interrupt Vectors, capable of managing multiple events with different priorities so as to be able to manage a hierarchical order of interruption requests.

Key Project Indicators KPI

The Key Performance Indicators (KPIs) are the indicators used to monitor the productivity of projects in general and of large IT projects in particular. The follow-up of the project is summarized in tables containing numerous indicators of progress of the various tasks and comparisons between what was "planned" and what was "achieved," in terms of time, budget and use of human and material resources. The management and manipulation of KPI tables is one of the main tasks for project management, with a special emphasis on tools used for empirical measuring and productivity control.

Linux, Unix

Unix is the first family of portable computer operating systems. Initially developed by a research group at AT&T and Bell Laboratories, which included Ken Thompson and Dennis Ritchie. Linux is a family of Unix-like operating systems, released under the GNU GPL license, under various possible distributions, having the common characteristic of using the Linux kernel as its core.

Neet

Acronym for Not in Education, Employment or Training, used to indicate people who are not involved in studies or training, do not have a job or who are looking for one.

Ordoliberalism

Ordoliberalism (Ordoliberalismus in German) is a liberal current of thought that appeared in Germany as early as 1930, according to which the state's mission is to create and maintain a legal framework that allows free competition between companies, ensuring that no obstacle stands in the way.

P2P

"Peer-to-peer" or "Peer2peer" or equal network, in computer science, is an expression that indicates a logical architecture of computer network in which the nodes are not hierarchized in the form of fixed clients or servers (clients and servers), but in the form of equivalent or equal nodes that can act both as client and server towards the other terminal nodes (hosts) of the network. By extension, the term is used for P2P movement. Here is a definition by M. Bauwens, founder of the P2P foundation:

> In reality, the term P2P has long been used to refer to a range of solutions, paradigms, approaches centered on co-design and co-creation, openness and freedom: that is, every decentralized, common, distributed, participated and equal approach in providing free and open solutions to common problems. Technology and technology platforms (and in particular software) are therefore only one of the many aspects of this movement, which in fact has no limits whatsoever: the long-term objective is to facilitate the emergence and consolidation of communities capable of playing a new role, a productive role traditionally reserved for companies according to the capitalist model of industrial production of goods and services. The model of equal production (peer production) is therefore in antithesis to

> neoliberalism but, inevitably, these processes on the one hand will transform and, on the other, adapt to today's society in search of that synthesis that perhaps represents the only way out of the historical problems that humanity is facing.

Proprietary software

Proprietary software, also called private, non-free, or closed source software, is copyrighted software whose license allows the beneficiary to use it under certain conditions and prevents others such as modification, sharing, study, redistribution or reverse engineering. The restrictions are imposed by the holder of the rights of economic exploitation (which are often large IT companies such as Apple and Microsoft). Proprietary software is software that does not allow the four software freedoms (described in the free software section of this glossary) to be exercised technically, legally or by other means. By analogy, the adjective "proprietary" is used in other techno-scientific fields to indicate a "privatization" of research results, such as those on the human genome.

Unconditional Basic Income

UBI is a primary (main) and unconditional income that is guaranteed to every adult citizen. According to the economist Carlo Vercellone:

> the proposal for an unconditional guaranteed social income independent of salaried work is the one that seems to us to embody a new stage of socialization of the economy. It is part of a social project and the demercantilization of the economy where the strengthening of collective rights linked to the social protections (pensions, health, unemployment benefits, etc.) would go hand in hand with the transition from a welfare model to a commonfare model. In fact, similarly to the theme of the common, the reflection on the proposal of an unconditional basic income (UBI) is increasingly debated as an alternative to the crisis of cognitive and financialized capitalism. However, most of the

> formulations of the UBI remain anchored to a conception that makes it a secondary income, related to the redistribution and the classical state management of the State-provision.
>
> In our approach, the UBI must, on the contrary, at the same time be thought of as an institution of the common and as a primary income for individuals, i.e. an income that results directly from production and not from redistribution. These two dimensions, primary income and institution of the common, are closely interwoven both on the level of production organization, as well as on that of the statute of the currency and the mode of distribution.
>
> A primary income, therefore, because the UBI proposal is based on a review and an extension of the concept of productive work, which must be taken into consideration in the light of two dimensions. The first one conceives productive work, according to the dominant tradition in the political economy, as the work that produces value and surplus value. This is the observation that today we are witnessing an important extension of working hours, outside the official working day, which are directly or indirectly involved in the formation of the value captured by companies. In this regard, UBI would correspond, in part, to the social remuneration of this increasingly collective dimension of a value-creating activity, which extends over all social times, giving rise to an enormous mass of unrecognized and unpaid work.
>
> Secondly, UBI, as primary income, presupposes and relaunches the development of the common itself. It would relaunch it to the extent that it would facilitate the exodus from paid work and the development of forms of cooperation based on coordination rules that are distinct from those of the public and the market. It presupposes this in so far as its establishment implies mechanisms for the re-socialization of currency and incomes which make the reproduction of the labour force independent of the circulation of money as capital." (Baronian & Vercellone, 2013)

Venture capital (Risk Capital)

Venture capital is the provision of risk capital by an investor to finance the start-up or growth of an activity in sectors with high development potential. A venture capital fund primarily invests financial capital in companies that are too risky for standard capital markets or bank lending. Often the same name is given to specially created funds, while the entities carrying out these operations are called venture capitalists.

Zoé and Bios

ζωή (zoé) means the principle, the essence of life, which belongs in common, indiscriminately, to the universality of all living beings and which has as its opposite concept non-life and not, as one might think, death, since this concerns the individual being who ceases, individually, to live. βίος (bios): indicates the conditions and the ways in which our life is carried out. Zoé is therefore the life that is in us and through which we live (qua vivimus), bios alludes to the way we live (quam vivimus), i.e. the modalities that characterize for example contemplative life, political life, etc. for which the Greek language uses precisely the term bios accompanied by a qualifying adjective.

BIBLIOGRAPHY

Agamben, G. (1998). *Homo Sacer: Sovereign Power and Bare Life.* Stanford: Stanford University Press.

Agamben, G. (2009). *What is an Apparatus? And other essays.* Stanford: Stanford University Press.

Alquati, R. (1961). “Organic Composition of Capital and Labor-Power at Olivetti,” *Viewpoint Magazine.* https://www.viewpointmag.com/2013/09/27/organic-composition-of-capital-and-labor-power-at-olivetti-1961/

Anderson, C. (2014). *Makers: The New Industrial Revolution.* London: Random House.

Annesi-Maesano, I., & Baïz, N. (2012). “Is the Asthma Epidemic Still Ascending?” *Clinics in Chester Medicine,* Volume 33 Issue 3: 419-429.

Armano, E., & Sacchetto, D. (2012). “La conricerca contro l’industrializzazione dell’umano,” in Infoaut (ed), *Romano Alquati. Immagini e percorsi soggettivi e collettivi di una ricerca.*

Arsenault, A., & Castells, M. (2008). “Switching Power: Rupert Murdoch and the Global of Media Politics,” *International Sociology* Volume 23 Issue: 4: 488-513.

Avelino, H. (2006). *The typology of Pame number systems and Berkeley.* http://linguistics.berkeley.edu/~avelino/Avelino_2006.pdf

Baronian, L., & Vercellone, C. (2013). “Moneta del comune e reddito sociale garantito,” *Uninomade 2.0.* http://www.uninomade.org/moneta-del-comune-e-reddito-sociale-garantito/intervista-a-michel-bauwens/

Bauwens, M., & Kostakis, V. (2013). *Network Society and Future Scenarios.* London: Palgrave Macmillan.

Bazzichelli, T. (2013). *Networked Disruption.* Copenhagen: Digital Aesthetics Research Center Press.

Benkler, Y. (2011). *The Penguin and the Leviatian: How Cooperation*

Triumphs over Self-Interest. New York: Random House.
Berardi, F. (2013). "Passaggio per Seoul," *Through Europe*. http://th-rough.eu/writers/bifoita/passaggio-seoul-2
Blondeau, O. (2007). *Devenir Media*. Paris: Editions Amsterdam.
Blumir, G. (1983). *Eroina. Storia e realtà scientifica. Diffusione in Italia. Manuale di autodifesa*. Milano: Feltrinelli.
Borges, L. (2000). *Fictions*. New York: Penguin Classics.
Bowyer, A. (2011). *Wealth without money*. York. http://reprap.org/wiki/Wealth_Without_Money
Braidotti, R. (2002). *Nomadic Subjects. Embodiment and Sexual Difference in Contemporary Feminist Theory, Second Edition*. New York: Columbia University Press.
Braidotti, R. (2013). *The Posthuman*. Cambridge: Polity Press.
Buterin, V. (2013). "Bootstrapping An Autonomous Decentralized Corporation, Part 2: Interacting With the World," *Bitcoinmagazine.com*.
Calvino, I. (1993). *If on a Winter's Night a Traveler*. New York: Penguin Random House.
Carr, N. (2012). *Net Delusion: The Dark Side of the Internet*. New York: Public Affairs.
Carr, N. (2008). *The Big Switch: Rewiring the World, from Edison to Google*. New York: Norton.
Carr, N. (2010, 6). "The Web Shatters Focus, Rewires Brains," *Wired*.
Case, A. (2008). "The Cell Phone and its Technosocial Sites of Engagement." http://caseorganic.com/thesis/
Cassavetes, N. (Director). (2002). *John Q* [Film].
Castells, M. (1998). *The Information Age: Economy, Society and Culture, vol. III, End of Millennium*. Malden: Wiley-Blackwell.
Castells, M. (2000). *The Rise of the Network Society: The Information Age: Economy, Society, and Culture Volume I*. Malden: Wiley-Blackwell.
Chamayou, G. (2015). *Drone Theory*. London: Penguin.
Chaplin, C. (Director). (1936). *Modern Times* [Film].
Chiesa, G. (Director). (2004). *Lavorare con lentezza* [Film].
Coleman, G. (2011). *Anonymous: From the Lulz to Collective Action*. http://mediacommons.futureofthebook.org/tne/pieces/anonymous-lulz-collective-action
Coleman, G. (2013). *Coding Freedom*. Princeton: Princeton University Press.

Cooper, M., & Waldby, C. (2014). *Clinical Labor: Tissue Donors and Research Subjects in the Global Bioeconomy*. Durham: Duke University Press.

Coriat, B. (1991). *Penser à l'envers*. Paris: Christian Burgois.

Cortázar, J. (1987). *Hopscotch: A Novel*. New York: Pantheon.

Damasio, A. (1995). *L'errore di Cartesio. Emozione, ragione e cervello umano*. Milano: Adelphi.

Damasio, A. (2006). *Descartes' Error: Emotion, Reason and the Human Brain*. Vintage.

Dawkins, R. (1990). *The Selfish Gene*. Oxford: Oxford University Press.

Day, R. (2005). *Gramsci is dead*. London: Pluto Press.

Delfanti, A. (2013). "Il lato oscuro della rete. Intervista a Michel Bauwens" in *Uninomade 2.0*. http://www.uninomade.org/

Deleuze, & Guattari. (1972). *Anti-Oedipus. Capitalism and Schizophrenia*. London: Penguin Random House.

Deleuze, G. (1992). "Postscript on the Societies of Control" in *October*, Vol 59 (Winter): 3-7.

Deleuze, G., & Guattari, F. (1987). *A Thousand Plateaus: Capitalism and Schizophrenia*. Minneapolis: University of Minnesota Press.

Des Isnards, & Zuber, T. (2011). *L'open space m'a tuer*. Paris: Ville de Poche.

Dyer-Witheford, N. (2013). *Red plenty platforms*. Available at https://culturemachine.net.

Erra, R. (2015). *Comment attaquer une place de Bourse en 30 minutes et comment se protéger? les nouveaux loups de Wall Street*. [Film] https://www.youtube.com/watch?v=GQ8ow2gcO3k

Faircoop. (2014). The Earth Cooperative for a Fair Economy. https://2017.fair.coop/the-earth-cooperative-for-a-fair-economy/

Fleischer, R. (Director). (1973). *Soylent Green* [Film].

Formenti, C. (2000). "Intervista a Carlo Formenti." http://www.autistici.org/operaismo/forment2/2_1.htm

Foucault, M. (2008). *The Birth of Biopolitics. Lectures at the Collège de France, 1978-79*. London: Palgrave Macmillan.

Foucault, M. (2007). *Security, Territory, Population*. London: Palgrave Macmillan.

Frears, S. (Director). (2013). *Philomena* [Film].

Freire, P. (1996). *The Pedagogy of the Oppressed*. New York: Penguin.

Fukuyama, F. (2006). *The End of History and the Last Man*. New

York: Free Press.
Fumagalli, A. (2015). *La vie mise au travail. Nouvelles formes du capitalisme cognitif.* Paris: Eterotopia France.
Fumagalli, A. (2015). *Le trasformazioni del sistema capitalistico e le nuove contraddizioni sociali.* https://www.academia.edu/8658153/le_trasformazioni_del_sistema_capitalistico_e_le_nuove_contraddizioni_sociali
Fumagalli, A., & Morini, C. (2011). "Life put to work: towards a theory of life-value," *ephemera*, vol. 10: 234-252.
Giannelli, G., & Griziotti, G. (2013). "Dumping on the world: e-waste 2.0" in OpenDemocracy. https://www.opendemocracy.net/gianluca-giannelli-giorgio-griziotti/dumping-on-world-e-waste-20
Gibson, W. (1986). *Neuromancer.* New York: Ace Books.
Gorz, A. (2010). *The Immaterial.* Calcutta: Seagull Books.
Gorz, A. (1982). *Ecology as Politics.* Montreal: Black Rose Books.
Gotardo, C. (Director). (2012). *O que se move* [Film].
Grifi, A. (Director). (1976). *Il festival del proletariato giovanile al Parco Lambro* [Film].
Griziotti, G. (2015). "Intervista su CIC e Faircoop: 12 domande a Enric Duran" in *La moneta del comune.* Alfabeta2-Derive Approdi.
Griziotti, G. (2013a). "Introduzione allo Speciale AlfaBioipermedia" in Alfabeta2 (29). Griziotti, G. (2013b). "Under the regime of precarity: bring your own device" in OpenDemocracy. https://www.opendemocracy.net/giorgio-griziotti/under-regime-of-precarity-bring-your-own-device
Griziotti, G. (2014). "Unblock the chain," *Effimera.* http://effimera.org/unblock-the-chain-i-processi-collaborativi-etecnologici-
Griziotti, G., & Vercellone, C. (2014). "Biorank vs Commoncoin," *Effimera.* http://effimera.org/biorank-vscommoncoin-
Grossman, V. S. (2006). *Life and Fate.* New York: New York Review Books.
Gubitosa, C. (2010). *Hacking Miracles.* Mumbai: Jaico Publishing House.
Haraway, D. (1991). "A Cyborg Manifesto: Science, Technology, and Socialist-Feminism in the Late Twentieth Century," in *Simians, Cyborgs and Women: The Reinvention of Nature.* Routledge.
Hardt, M., & Negri, A. (2001). *Empire.* Cambridge: Harvard

University Press.

Hardt, M., & Negri, A. (2005). *Multitude: War and Democracy in the Age of Empire*. London: Penguin.

Hardt, M., & Negri, A. (2011). *Commonwealth*. Cambridge: Harvard University Press.

Harvey, D. (2013). *Rebel Cities: From the Right to the City to the Urban Revolution*. London: Verso.

Himanen, P. (2007). *The Hacker Ethic and the Spirit of the Information Age*. New York: Random House.

Hobsbawm, E. (1994). *The Age of Extremes: The Short Twentieth Century, 1914-1991*. London: Vintage Books.

Hobsbawm, E. (2012). "Lenin and the 'Aristocracy of Labor,'" *Monthly Review*, Vol 64, Issue 07.

Holbrook, H. (1982). "The experiential Aspects of Consumption: Consumer Fantasy, Feelings and Fun" in *Journal of Consumer Research*, 9 (2).

Huws, U. (2010). "Expression and expropriation: The dialectics of autonomy and control in creative labour," *ephemera* Volume 10 Number 3: 504-521.

Iacoboni, M. (2008). *Mirroring People: The New Science of How We Connect with Others*. New York: Farrar, Straus and Giroux.

Ippolita. (2005). "Open non è Free. Comunità digitali tra etica hacker e mercato globale." *Eleuthera*.

Ippolita. (2012). "Nell'Acquario di Facebook." Ippolita.net.

Jonze, S. (Director). (2013). *Her* [Film].

Kamata, S. (1982). *Japan in the Passing Lane: An Insider's Account of Life in a Japanese Auto Factory*. New York: Pantheon Books.

Kandel, E. (2007). *In Search of Memory. The Emergence of a New Science of Mind*. New York: W. W. Norton & Company

Kerouac, J. (2007). *On the Road: The Original Scroll*. Penguin Random House.

Kleiner, D. (2010). *The Telekommunist Manifesto*. Amsterdam: Institute of Network Cultures.

Kore-eda, H. (Director). (2013). *Soshite chichi ni naru*. [Film].

Lallement, M. (2015). *L'Age du faire*. Paris: Seuil.

Lazzarato, M. (2004). *Les révolutions du Capitalisme*. Paris: Les empêcheurs de penser en rond.

Lazzarato, M. (2014). *Signs and Machines Capitalism and the Production of Subjectivity*. Los Angeles: Semiotexte.

Lebert, D., & Vercellone, C. (2010). "Capitalisme cognitif et capitalisme mafieux," Uninomade 2.0. http://uninomade.org/quels-rapports-etablissez-vous-entre-capitalisme-cognitif-et-capitalismemafieux/

Lovink, G. (2012). *Networks Without a Cause: A Critique of Social Media.* Cambridge: Polity.

Mackey, & Sisodia, R. (2013). Conscious Capitalism Is Not an Oxymoron. *Harvard Business Review.* http://blogs.hbr.org/2013/01/cultivating-a-higher-conscious/

Malet, J.-B. (2013). *En Amazonie. Un infiltrato nel "migliore dei mondi."* Paris: Kogoi Edizioni.

Marazzi, C. (2011). *Capital and Affects. The Politics of the Language Economy.* Los Angeles: Semiotexte.

Marazzi, C. (2008). *Capital and Language: From the New Economy to the War Economy.* Los Angeles: Semiotexte.

Marazzi, C. (2010). *Il comunismo del capitale.* Verona: Ombre corte.

Marazzi, C. (2015). "La nemesi storica del capitale" in Effimera.org. http://effimera.org/la-nemesi-storica-del-capitale-intervista-a-christian-marazzi-di-gigiroggero/

Marx, K. (1973). *Gundrisse. Foundations of the Critique of Political Economy.* London: Penguin Books.

Mass, G., & Kaufmann, J. (Director). (2012). *Two Lives* (Zwei Leben) [Film].

Mazlish, B. (1993). *Fourth discontinuity.* New Haven: Yale University Press.

Mbembe, A. (2003). "Necropolitics," *Public Culture* Vol 15 (1): 11-40.

Mc Luhan, M. (1991). *La galassia Gutenberg. Nascita dell'uomo tipografico.* Rome: Armando Editore.

Mc Gimsey, J. F. (2012). "From Parchment to Picture: On the Road," *Paris Lit Up.* http://parislitup.com/from-parchment-to-picture-on-the-road/

Mehri, D. (2006). "The Darker Side of Lean: An Insider's Perspective on the Realities of the Toyota Production System," *Academy of Management Perspectives* Volume 2 Number 2: 21-42.

Monnier, J.-M., & Vercellone, C. (2013). "Le financement du revenu social garanti comme revenu primaire," *Muovements*: 44-53.

Moore, M. (Director). (2007). *Sicko* [Film].

Morini, C. (2015). "Materiali viventi nella catena del valore," *Il Manifesto.*

Morozov, E. (2012). *Contro Steve Jobs*. Turin: Codice Edizioni.
Morozov, E. (2013). *To Save Everything, Click Here: The Folly of Technological Solutionism*. New York: PublicAffairs.
Munster, A. (2011). "Nerves of Data: The Neurological Turn In/ Against Networked Media" in Computational Culture. http:// computationalculture.net/nerves-of-data/
Negri, T. (2007). *Dall'operaio massa all'operaio sociale. Intervista sull'operaismo di Toni Negri*. Verona: Ombre Corte.
Negri, T., & Vercellone, C. (2008). "Le rapport capital-travail dans le capitalisme cognitif" in *Multitudes* (32).
Niccol, A. (Director). (2011). *In Time* [Film].
Niccol, A. (Director). (2015). *Good Kill* [Film].
OECD. (2005). *Oslo Manual. The Measurement of Scientific and Technological Activities*. OCDE. https://www.oecd.org/sti/ inno/2367580.pdf
Paolini, M. (2012). *Ausmerzen*. Torino: Einaudi.
Perec, G. (2009). *Life: A User's Manual*. David R. Boston: Godine Publishers.
Petri, E. (Director). (1971). *The Working Class Goes to Heaven* [Film].
Philopat, M. (2006). *Costretti a Sanguinare*. Torino: Einaudi.
Piketty, T. (2014). *Capital in the Twenty-First Century*. Cambridge: Harvard University Press.
Poitras, L. (Director). (2014). *Citizenfour* [Film].
Rancière, J. (1991). *The Ignorant Schoolmaster: Five Lessons in Intellectual Emancipation*. Stanford: Stanford University Press.
Raymond, B. Y. (2001). *The Cathedral & the Bazaar*. Boston: O'Reilly.
Reed, J. (1919). *Ten Days that Shook the World*. New York: Boni & Liveright.
Rifkin, J. (2015). *The Zero Marginal Cost Society: The Internet of Things, the Collaborative Commons, and the Eclipse of Capitalism*. New York: St. Martin's.
Roio, D. J. (2013). "Bitcoin, la fine del tabù della moneta" in Effimera.org. http://effimera.org/ bitcoin-la-fine-del-tabu-della-moneta-di-denis-jaromil-roio/
Sassen, S. (2015). *Losing Control? Sovereignty in the Age of Globalization*. New York: Columbia University Press.
Scott, R. (Director). (1982). *Blade Runner* [Film].
Simondon, G. (2016). *On the Mode of Existence of Technical Objects*.

Minneapolis: Univocal Publishing.
Simondon, G. (2006). "Technical Mentality," *Parrhesia* Issue 7. Available at www.parrhesiajournal.org.
Stallmann, R. (2002). *Free Software, Free Society: Selected Essays of Richard M. Stallman*. GNU Press.
Stephenson, N. (1999). *In the Beginning was the Command Line*. New York: HarperCollins.
Stiegler, B. (2013). "Réinventer un rapport au temps," *Le Monde*. Paris. http://
reseaux.blog.lemonde.fr/2013/10/03/ reinventer-rapport-temps-bernard-stiegler
Terranova, T. (2012). "Attention, Economy and the brain," *Culture Machine*, 13. http://www.culturemachine.net/index.php/cm/article/view/465/484
Terranova, T. (2013). "Capitalismo cognitivo e vite neurali" in Alfabeta2. Rome.
Terranova, T. (2014). "Red stack attack! Algoritmi, capitale e automazione del comune" in Effimera.org. http://effimera.org/red-stack-attack-algoritmi-capitale-e-automazione-del-comune-di-tiziana-terranova/
Terranova, T. (2015). "Securing the social: Foucault and Social Networks." https://www.academia.edu/11258389/Securing_the_social_Foucault_and_Social_Networks
Tyldum, M. (Director). (2014). Imitation Game [Film].
Vecchi, B. (2012). "Il disincanto di Ippolita e lo sboom di Facebook a Wall Street" in
Uninomade.org. http://www.uninomade.org/ disincanto-ippolita-e-sboom-facebook/
Venter, C. (2007). *A life decoded*. New York: Penguin Books.
Vercellone, C. (2002). "Les politiques du développement à l'heure du capitalisme" in *Multitudes*, 14. Paris
Vercellone, C. (2006). *Capitalismo cognitivo*. Manifestolibri.
Vercellone, C. (2007). "From Formal Subsumption to General Intellect" in *Historical Materialism* (15 (1)): 13-36.
Vercellone, C. (2015). "Reddito di cittadinanza: quale spazio per il lavoro?" in Euronomade. http://www.euronomade.info/?p=5339
Vercellone, C., & altri. (2015). *Managing the commons in the knowledge economy*.
European Union's Seventh Framework Programme for research,

technological development and demonstration project, D-CENT Decentralised Citizens engagement Technologies.

Virilio, P. (1996). *Cybermonde, la politique du pire.* Textuel.

Weber, M. (1930). The Protestant Ethic and the Spirit of Capitalism. London: Unwin Hyman.

Weir, P. (Director). (1998). *Truman Show* [Film].

Williams, & Srnicek, N. (2013). "#ACCELERATE MANIFESTO for an Accelerationist Politics" in Critical Legal Thinking. http://criticallegalthinking.com/2013/05/14/accelerate-manifesto-for-an-accelerationist-politics/

Yee, N. (2007). "Motivations of Play in Online Games," *Journal of Cyberpsychology and Behavior*, 9, 772-775.

Yuill, S. (2008). *Interrupt.* Software Studies.

MINOR COMPOSITIONS

Other titles in the series:

Precarious Rhapsody – Franco "Bifo" Berardi
Imaginal Machines – Stevphen Shukaitis
New Lines of Alliance, New Spaces of Liberty – Felix Guattari and Antonio Negri
The Occupation Cookbook
User's Guide to ~~Demanding~~ the Impossible – Laboratory of Insurrectionary Imagination
Spectacular Capitalism – Richard Gilman-Opalsky
Markets Not Capitalism Ed. Gary Chartier & Charles W. Johnson
Revolutions in Reverse – David Graeber
Undressing the Academy – University for Strategic Optimism
Communization & its Discontents – Ed. Benjamin Noys
19 & 20 – Colectivo Situaciones
El Martillo – Eclectic Electric Collective
Occupy everything! Reflections on 'why its kicking off everywhere' – Ed. Alessio Lunghi and Seth Wheeler
Punkademics – Ed. Zach Furness
[Open] Utopia – Thomas Moore & Stephen Duncombe
Contract & Contagion – Angela Mitropoulos
Squatting in Europe – Squatting Europe Kollective
Artpolitik: Social Anarchist Aesthetics in an Age of Fragmentation – Neala Schleuning
The Undercommons – Fred Moten & Stefano Harney
Nanopolitics Handbook – Nanopolitics Group
Precarious Communism: Manifest Mutations, Manifesto Detourned

– Richard Gilman-Opalsky
FutureChe – John Gruntfest & Richard Gilman-Opalsky
Lives of the Orange Men – Major Waldemar Fydrich
State in Time – Irwin
Crisis to Insurrection – Mikkel Bolt Rasmussen
Occupation Culture – Alan W. Moore
Crisis to Insurrection – Mikkel Bolt Rasmussen
Gee Vaucher. Introspective – Ed. Stevphen Shukaitis
The Aesthetic of Our Anger – Ed. Mike Dines & Matthew Worley
The Way Out – Kasper Opstrup
Situating Ourselves in Displacement – Ed. Marc Herbst et al.
Organization after Social Media – Geert Lovink and Ned Rossiter
Don't Network – Marc James Léger

www.ingramcontent.com/pod-product-compliance
Lightning Source LLC
Jackson TN
JSHW021706120925
90923JS00001B/28
* 9 7 8 1 5 7 0 2 7 3 4 2 1 *